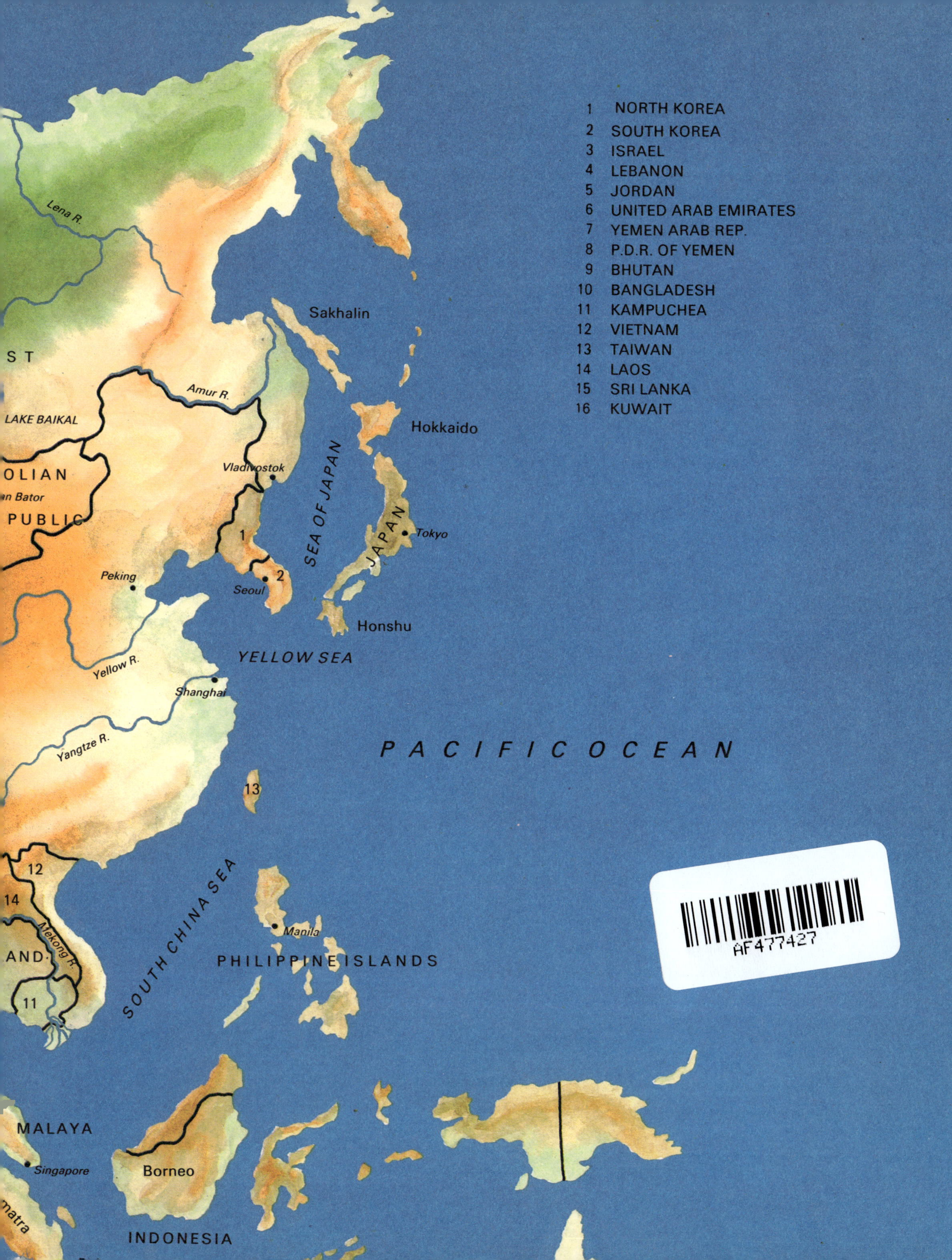

1   NORTH KOREA
2   SOUTH KOREA
3   ISRAEL
4   LEBANON
5   JORDAN
6   UNITED ARAB EMIRATES
7   YEMEN ARAB REP.
8   P.D.R. OF YEMEN
9   BHUTAN
10   BANGLADESH
11   KAMPUCHEA
12   VIETNAM
13   TAIWAN
14   LAOS
15   SRI LANKA
16   KUWAIT
Lena R.
LAKE BAIKAL
Amur R.
Sakhalin
ST
OLIAN
PUBLIC
an Bator
Vladivostok
Hokkaido
SEA OF JAPAN
JAPAN
Peking
Tokyo
Seoul
Honshu
Yellow R.
YELLOW SEA
Shanghai
Yangtze R.
PACIFIC OCEAN
13
12
14
Mekong R.
SOUTH CHINA SEA
AND
Manila
11
PHILIPPINE ISLANDS
MALAYA
Singapore
Borneo
matra
INDONESIA

ANIMALS OF ASIA

ANIMALS OF ASIA

DR JIŘÍ FELIX

ILLUSTRATED BY
KVĚTOSLAV HÍSEK, JAROSLAV KNOTEK
AND LIBUŠE KNOTKOVÁ

HAMLYN

LONDON • NEW YORK • SYDNEY • TORONTO

Designed and produced by Artia
for The Hamlyn Publishing Group Limited
London • New York • Sydney • Toronto
Astronaut House, Feltham, Middlesex, England
© Artia, Prague 1983
© This edition by
The Hamlyn Publishing Group Limited
Translated by Dana Hábová
Graphic design by Vladimír Šmerda
Photographs: ČTK (p.240), Werner Forman (p.192),
Ivan Koudelka (p.262), Oldřich Mazůrek (p.16,152),
Stanislav Wieser (p.102)

ISBN 0 600 36642 1
Printed in Czechoslovakia by Svoboda, Prague
1/12/06/ 51-01

CONTENTS

INTRODUCTION

Asia is the largest continent. Its name comes from the Assyrian word 'asu', which means 'region of the East'. Together with Europe, it forms a huge natural land mass; Europe is separated from Asia by the Ural Mountains. Asia also includes many islands: there are three groups of islands in the Arctic Ocean in the north, in the east stretch the large islands of Sakhalin, Taiwan, the Japanese Islands, the Philippines, the Indonesian Islands, Sri Lanka and countless other smaller islands and islets.

The north of Asia has a very cold, arctic climate and large glaciers, while the south has hot tropical weather. A temperature of 78° C below zero has been measured in Siberia and a temperature of 50° C has been recorded in Baghdad. Nowhere in the world exists such an extreme difference between the highest and lowest points: Lake Baikal, with its depth of 1 742 m, is the deepest lake in the world, and Mount Everest in the Himalayas, reaching a height of over 8 840 m, is the highest peak.

Asia has examples of all the vegetation zones in the world. Tundra, a cold, treeless plain, covers a belt as much as 600 km wide in the north, ranging from the Kola Peninsula to Chukotka. In some areas, the tundra passes from lowlands to mountains. Despite a brief growing period, the countryside is teeming with life. The lakes and rivers are crystal clear. The tundra is an important source of oxygen for the inland regions of Asia. Arctic winds, reaching the remote interior of the continent, bring fresh cooled air rich in oxygen. In the south, the tundra gradually changes into an extensive zone of coniferous forest — the taiga. Steppes (enormous grasslands) and deserts are found farther in the south, and these then give way to the subtropical and tropical zones. Most of the regions of southern, south-eastern and eastern Asia are situated in the monsoon zone with regular seasons of rain and drought. Each zone has its characteristic flora and fauna.

Although Asia has immense uninhabited areas of desert, tundra, mountains and forests, it is home for more than half of the world's population. People are concentrated in favourable locations such as river deltas, fertile valleys and coastal regions. Japan, eastern China, Bengal and Java are the most densely populated areas in Asia.

Asia is the home of many wild animals and some species are found only in this continent. Such animals include the legendary Giant Panda, the Orang-utan and the Tiger. Some endemic, that is locally occurring, species were ancestors of our most important domestic animals. An ancestor of many domesticated cattle breeds, the Wild Auroch, was probably first domesticated in the territory of present-day Turkestan and Uzbekistan. The Zebu breeds originated from the extinct Indian Aurochs. The Water Buffalo is one of the most important domestic animals in southern and south-eastern Asia. It is a descendant of the Arni or Wild Water Buffalo which lives in the wild state from India to Kalimantan. The Yak is kept by the inhabitants of the high Tibetan mountains, and its wild ancestor can still be found on mountain plateaux. One of the horse's ancestors once grazed in the Asiatic steppes. The Asiatic Wild Goat is related to domestic goats and the domestic sheep comes from western Asia. The most important domestic bird, the chicken, found almost all over the world, is a descendant of the Red Jungle Fowl from southern Asia. Other tamed or domesticated wild animals, such as elephants, camels and reindeer, are used to help people with heavy work.

There are now more than three hundred large national parks and wildlife sanctuaries in Asia, preserving the original flora and fauna and saving many species of plants and animals from extinction. However, it is still vitally important to enlarge the protected territories and to support rare animal species living outside them. Many species of Asiatic animals have been destroyed by man. Steller's Sea Cow was hunted out of existence along the north-eastern coast by 1768. Other species, such as Père David's Deer, whose original habitat is unknown, are now only found in captivity. The wild Przewalski's Horse has probably suffered the same fate; it was last reported in the wild in 1960. No wild horse has been seen since, and it is now probably found only in captivity. The Manchurian Pheasant is a bird representative of the extinct Asiatic fauna; it was resident in the Shansi Mountains north of Peking. This list covers only a few cases of the exterminated or adversely affected Asiatic animals. Only consistent protection of the wildlife and its habitats can prevent other species from being annihilated in the uneven war with man. This requires an enormous effort both by conservationists and the governments of all Asiatic countries.

THE
TROPICAL
PRIMARY
FORESTS

Asia is the largest of the continents. It occupies an enormous territory extending from the Arctic Ocean in the north to the Indian Ocean in the south, and from the Mediterranean Sea in the west to the Pacific Ocean in the east. This vast land encompasses the severe cold of the arctic north and the extreme heat of the tropics in the south.

Extensive areas of tropical primary forest, mainly rainforest, are found throughout southern and south-eastern Asia including the islands of Indonesia and Eastern Malaysia. These comprise the Greater Sunda Islands of Sumatra, Borneo and Java, and the Lesser Sunda Islands. Sulawesi and the Philippines also form part of tropical Asia. The tropical primary forests are predominantly deciduous, though mixed woodland and coniferous forests can be found in the mountains. In the extreme south of Asia, tropical forests occur at heights of up to 2 000 m. Forests in low-lying situations and on the hills contain an immense variety of species. In many areas, fifty different species can be found among a hundred trees on a single site, and of them twenty-five belong to different families. Although trees are in bloom all year round, for the most part the colourful petals are invisible in the ocean of greenery. Only when a few varieties of trees shed their leaves, and before new foliage appears, do the naked branches shimmer with an avalanche of bright blossoms. Such species include silk-cotton trees of the genus *Bombax,* which are found on the Indian subcontinent and in south-east Asia. These enormous trees can grow to a height of 40 m, their crowns forming huge umbrellas above the surrounding vegetation. The Silk-cotton Tree *(Bombax malabaricum)* is most commonly encountered, especially in south-east Asia. In spring, before the leaves begin to bud on the smooth bark, the large, scarlet blossoms of this tree shine like lights. The 30 or so species of coral trees of the genus *Erythrina* also have magnificent red blossoms. In particular *Erythrina indica* is frequently seen among the forest vegetation, and it is also grown in plantations to provide shade in the heat of the day for coffee shrubs and other cultivated plants. Trees of this genus are of great importance to many bird and insect species, for the flowers provide sweet nectar for hundreds of birds and for thousands of butterflies. Asian primary forests also contain huge numbers of fig trees of the genus *Ficus,* their spreading crowns sheltering nesting birds, and the ripe fruits providing food in abundance. The Banyan *(Ficus benghalensis)* and *Ficus glomerata* are the most common. Some species of fig grow to enormous size and have trunks which can scarcely be encircled by four men. The

Weeping Fig *(Ficus benjamina)* of south-east Asia is one of these giants, characterized by its long aerial roots. Such trees often exceed 30 m in height, and many animals find shelter among the dense branches which are covered with large thick leaves. Birds of prey often nest in the tops of these trees, using the high branches as vantage points from which to view their hunting grounds.

Most important to man are the species which yield valuable timber. In particular, teak and other related species provide extremely hard, heavy and durable wood, often used in the construction of bridges which are required to last for many years in unfavourable tropical or maritime climates. This timber is often more suitable than iron structures, for it is less susceptible to the effects of a humid or saline environment. The timber from these trees is frequently used in shipbuilding for dockyard equipment. Teak *(Tectona grandis)* of India and Malaysia is light yellowish-brown in colour, and is both hard and highly resistant to water. It is a major item of export, and it is extensively cultivated in many places, particularly Malaysia. The medium-sized Ironwood *(Mesua ferrea)* of India, which is about 25 m tall, is also well-known for its dark red timber. The scented flowers of this tree are used in perfumes. Trees of the genus *Calophyllum,* such as the Indian *Calophyllum tomentosum,* are universally exploited, especially for the so-called Indian mahogany and for balsam obtained from the bark. Indian satinwood trees of the genus *Chloroxylon* yield very hard and beautifully mottled, yellow wood. Malaysia abounds in trees of the genus *Nephelium,* which have very durable timber, and these trees are also found in other regions of south-east Asia. Ebony, which is derived from several species of the genus *Diospyros,* is a particularly valuable timber. The Indian trees yield jet black wood, those of Sri Lanka have brown and black mottling and other species are pinkish, with dark stripes.

Species of the genus *Bauhinia* are an important component of Asian tropical forests. There are over 150 species and they exist both as trees and as massive, thick-trunked lianas climbing on the trunks of other trees. Asian tropical forests are the home of trees of the genus *Myristica.* The fruit of the Nutmeg Tree *(Myristica fragrans)* provides the spices nutmeg and mace. The Durian Tree *(Durio zibethinus)* is well-known from travel and adventure books. Durian trees are found in Malaysia and Indonesia and are often grown in plantations for their elliptical fruits. These are 20—30 cm long, and have hard, greenish rind. The seeds are coated with soft pink pulp with the consistency of custard. The fruits are extremely popular locally in spite of their unpleasant smell, like hydrogen sulphide and garlic, which is especially repulsive to Europeans. However, even for them, durian fruits can become a delicacy.

Palms also rank among the most important tropical species. On the Indian subcontinent and in south-eastern Asia, it is mainly the 20 m tall palms of the genus *Borassus* which are exploited. The pulp of the large round fruits is edible and the juice is processed to make palm wine and palm sugar. The leaves are used to make fans and a variety of other objects. The Betel Palm *(Areca catechu)* has a particular significance, especially in south-east Asia. This palm grows in abundance along the edges of primary forests. The slender trunks, 10—15 m tall, are topped with a few large leaves. The fruits, which are 5 cm long, are called betel nuts. Unripe seeds contain tannins and red pigments. Local inhabitants cut up or crush the seeds, smear them with lime and wrap them in leaves of the Betel Pepper *(Piper betel),* which have a tea-like scent but a sharp taste. The chewed leaves stimulate digestion and relieve fatigue, but they make the saliva red, and permanently blacken the teeth. A European is usually startled when a local girl smiles and reveals her shining black teeth. Lianas are a characteristic component of the tropical primary forests of Asia, especially the 200 or so species of the genus *Calamus,* commonly called rotang. They form scrubland thickets on the ground, and send out thin stems, 1—3 cm across, which climb up the trees. These stems are very strong and look like ropes many metres long. When they reach the top of a tree, they climb down again, often forming curtain-like walls which can only be penetrated by means of a sharp machete. Lianas are used locally for roofing huts, fastening fencing planks together, and securing boats, for they can be picked up only a stone's throw from where they are needed. Rotang fruits which are the size of hazel nuts, are a source of food for many birds, especially large species of parrots. Lianas of the genus *Gnetum* are common in Borneo and Malaysia. They have clusters of tiny blossoms and many small fruits which are eaten by birds.

The most important species native to the Indian subcontinent, China and south-eastern Asia, are the trees and shrubs of the genus *Citrus,* which bear the lemons and oranges common in almost every household all year long. Asian citrus trees have been distributed throughout the tropical and subtropical regions of the world and are now cultivated in vast plantations in

America, Africa, Australia and southern Europe. Asian lemons appeared in Persia and Egypt around 2 000 B. C. The Asian tropics are the home of another fruit which is known world-wide. The banana of the genus *Musa* is an ancient Indian crop. It is the largest of the plants in which the part growing above the ground withers when the fruits have ripened. It regenerates annually from its rootstalks. The bases of the leafstalks of the many leaves grow in dense spirals giving the impression of a solid trunk several metres tall. They each taper into a leafstalk with an enormous leaf blade, 2—4 m long and 60—100 cm wide. In due course, a flowerstalk grows up the middle of the leafstalk tube, and a pink to dark mauve inflorescence appears in the centre of the leaves. Wild bananas have bright red flowers, while cultivated species usually have violet ones. Banana plants do not just yield tasty fruits. The whole of the green part can be used. Shredded leafstalks are fed to pigs and the massive leaves are used for roofing huts. By combing the leafstalks of the related Abaca *(Musa textilis)* of the Philippines, fibres up to 2 m long are obtained. These silky, shining, solid fibres are processed into cordage called Manila hemp. Nowadays bananas are grown in humid tropical areas in every continent, being found in huge plantations as well as in small back gardens. The fruits can be harvested throughout the year and often constitute a major part of the diet of the local people.

Other plants of tropical India are used to produce dyes. In particular Indigo *(Indigofera tinctoria)* yields the well-known blue dye much used in the past and still an important export item.

Bamboo is another characteristic plant of the Asian tropics. It is a woody perennial grass which can reach a height of up to 40 m and which measures 30 cm in diameter. It can grow at the remarkable speed of one metre a day. It is found in forests, beside water, in mountains and in parks and gardens, and in some areas it forms vast bamboo thickets. Huge tufts of bamboo stalks, strengthened by joints which sprout lateral twigs, can be seen everywhere. There are scores of species of bamboo in Asia, the largest, *Dendrocalamus giganteus,* being found in the south-east. *Bambusa arundinacea* is a smaller species with yellowish stalks and thorny twigs, and like *Bambusa vulgaris* is widely distributed. For many Asians, bamboo is the most important crop, for this plant is indispensable to dwellers in both town and country, to those who live inland and to people of the coasts. Hundreds of tonnes of hollow bamboo stalks are processed every day in special factories and in the homes. Bamboo is used for furniture, wall-covering,

13

baskets, fences, poultry houses, rafts, decorations and even for pipelines to convey cool, clear water from mountain springs to villages in the valleys. By cutting out the joints and connecting long stems, a primitive pipeline can be made which stretches several hundred metres. Young bamboo shoots are a nourishing delicacy, and bamboo leaves feed many wild and domestic animals. Bamboo thickets provide ideal shelter for innumerable animals, ranging from tiny rodents to large sambar deer.

Darkness reigns continually in the heart of a tropical rainforest even in the middle of the day, for the sun's rays cannot penetrate the thick tangle of branches, lianas and epiphytic plants which in some places virtually hide the trunks of the huge forest trees. Epiphytes include not only orchids, which are mainly represented by the genus *Dendrobium* as well as by hundreds of other species in the Indonesian region, but also mosses and ferns. Arboreal ferns reach gigantic dimensions in humid areas, those of the genus *Angiopteris* which grow along the edges of forests having leaves 2—4 m long. The primary forests of Malaysia and of the Sunda Islands also feature epiphytic, myrmecophilous plants. These are small shrubs with tiny whitish or yellowish flowers, which provide shelter for ants. Holes in the tubers of the Javan species *Hydnophytum montanum* and of other species of the genus *Myrmecodia* serve as nests for a very aggressive species of ant, which rushes out at the slightest provocation, and attacks any intruder. In this way, the ants protect their nesting plant from beetles or caterpillars which might devour the leaves. Myrmecophilous trees of the genus *Batschia* are found in India. The ants nest in hollow sections of flowering branches, crawling through longways crevices to get inside.

Remarkable creeping insectivorous plants, called pitcher plants, of the genus *Nepenthes*, grow in the Indo-Malaysian region. The leaves have lidded pitchers 10—30 cm long on the tips, and their mottled, multicoloured appearance attracts insects to them. Each pitcher contains digestive juices that decompose insects which are trapped inside. In this way the pitcher plant feeds on the insects. These plants are of no danger to man, though adventure stories have often told of mysterious carnivorous plants trapping and devouring unfortunate travellers.

However, Asian primary forests can prove dangerous to those who visit them, not on account of tigers and leopards, as many believe, but because of the menace of tiny invertebrates. These include mosquitoes, which transmit the dreaded disease, malaria. There are many places in the

humid Asian tropics where clouds of these stinging insects await their prey. It is impossible to go hunting after sunset without the protection of a tropical helmet equipped with a veil covering the face and neck. A peaceful night can be spent in bed only under the cover of a mosquito net. Tiny leeches which live on leaves of trees and shrubs in the rainforests are another danger. An environment where the air is totally saturated with water, and where steam rising to the treetops condenses into rivulets which flow down to the ground again, provides ideal haunts for land leeches. Hundreds of these parasites lie in wait for the hapless traveller and the slightest disturbance of a branch can trigger off an attack. In the forest darkness they attach themselves to shirts or trousers, and then crawl on the skin of their involuntary hosts. Although this blood-thirsty parasite is gnawing at the skin, it is not usually noticed. However, the leech exudes a substance into the wound which stops the blood from clotting, and this results in the victims finding with horror, after the leech has left, that rivulets of blood are flowing down their arms or legs. Even a large number of leeches could not cause fatal bleeding, but the wounds can become dangerously infected. Local people smear leech wounds with lime, but their bodies are often covered with sores. A long-sleeved, buttoned-up shirt and long linen trousers tucked into the boots provide the best protection, and many women from villages situated in the rainforests wear long, tight-fitting trousers.

The forests also abound in beetles, butterflies, wasps, dragonflies, ants and termites. Asian termites do not build such huge termitaria as do their African or Australian counterparts, but settle in trees where they build spherical nests on the branches.

The primary forests of the Sunda Islands contain many unusual animals. Sumatra and Borneo — which comprises Kalimantan and Sarawak — are the home of the only Asian ape, the Orang-utan, which is described in many folk legends.

The flora of the rainforests is also unique. The largest flower, that of the parasitic plant Rafflesia *(Rafflesia arnoldii)* is found in Sumatra. This extraordinary plant lives on the roots of lianas of the genus *Cissus,* and its red-spotted flowers are a metre in diameter. The magnificent flower smells of carrion, and pollination is carried out by flies which buzz noisily in huge swarms around the plants.

PHILIPPINE FLYING LEMUR
Cynocephalus volans

The Philippine Flying Lemur inhabits tropical rain-forests in the Philippines. It reaches a length of 70 cm including the tail, which is about 25 cm long. This nocturnal animal can weigh in excess of 1.5 kg. The lemur sleeps throughout the day, hanging from a branch among the thick crowns of the trees, out of sight of its enemies. It wakes at dusk and hops slowly along the branches, or hangs beneath them as it climbs, gripping with its powerful claws. It often reaches other trees by making long gliding jumps. It has a wide membrane which stretches from the tip of the front limbs to the tip of the hind limbs and the tail. The furry membrane is folded under the front limbs when the animal is climbing or sitting so that it does not get caught in the branches. When it wants to jump, it opens up the membrane, springs away from its branch, and glides to another, travelling as far as 140 m while descending through only 10—13 m. It seeks its food among the treetops, feeding solely on vegetable material — leaves, flowers and fruits. In the courtship period the male finds a mate and stays with her for a few days, but for most of the year, these lemurs live singly. After a gestation period of 2 months, the female bears a single young, or occasionally twins. At first the young lemur clings to its mother's abdominal fur with claws. She carries it with her for some time, but later places it in a permanent nest in a dense treetop.

MOONRAT
Echinosorex gymnurus

The Moonrat is indigenous to Thailand, Malaysia, Sumatra and Borneo. It weighs up to 14 kg and reaches a length of 65 cm including the long, sparsely scaled tail, which is about 20 cm long. The body is covered with very long hairs. During the day this insectivore shelters in hollow tree stumps, under roots, or in crevices. It comes out to forage at dusk, combing the ground, especially on river banks, in its search for insects and their larvae, worms and fallen fruit. It also hunts in shallow water, catching small fishes and frogs. The Moonrat is a solitary animal, pairs forming only in the breeding season. The female gives birth to twins in a sheltered hollow. The whole of tropical Asia hosts scores of other insectivorous species.

INDIAN FLYING FOX
Pteropus giganteus

The Indian Flying Fox is distributed throughout Sri Lanka and the Indian subcontinent as far as the southern slopes of the Himalayas. The body of this fruit bat is about 30 cm long, and its wingspan can exceed 120 cm. It is gregarious, living in large colonies of hundreds, or even thousands. In the daytime the entire colony rests in the same huge tree. The bats hang upside down on horizontal branches, wrapped in their flying membranes. The crown of the tree is packed with sleeping flying foxes, the urine and excrement of which often destroy all the foliage. Activity begins at dusk, when the animals start climbing up the roosting tree, making screeching noises as they go. The enormous flock then takes off and moves like a black cloud to the forests. Here the bats forage for sweet juicy fruit, often causing considerable damage to gardens and fruit plantations. In the morning they return to their roosting site, keeping a fixed distance between each other. The dominant individuals settle on the highest branches, while the subordinate ones occupy the lowest layers of the tree and consequently suffer most from the urine and droppings. This species does not have a definite breeding season. After a gestation period of about 180 days the female gives

Philippine Flying Lemur

Moonrat

birth to a single young. The sparsely-furred infant clutches its mother's fur with its sharp claws and stays with her for four months. After this time it becomes independent and is capable of flight. A young flying fox is fully grown when it is one year old. Local people hunt Indian flying foxes for their tasty meat.

INDIAN SHORT-NOSED FRUIT BAT
Cynopterus sphinx

The Indian Short-nosed Fruit Bat is a common inhabitant of woodland in India, Sri Lanka and south-east Asia, including the Malay Peninsula. It measures up to 12.5 cm, and is found in abundance from the lowlands up to heights of about 1 850 m in the mountains. It lives in groups of 6—12, sheltering by day in caves or tree hollows, or under roofs or beneath large palm leaves. After sunset the flocks come out to forage for ripe figs, mangoes and bananas. Banana plantations often suffer considerable damage. Fruit bats sometimes lick nectar from flowering trees, especially from the Sausage Tree *(Oroxylum indicum)*. They are extremely active and are capable of sustained flight, often covering over 100 km in a single night. Usually in March or April, but now and then in other months, the female gives birth to a single young. It attaches itself to her body and is carried along as she hunts. In northern Thailand local inhabitants catch these bats in large numbers and sell them in the markets, for the meat of fruit bats is believed to cure all manner of diseases and also give strength to those who cat it.

INDIAN FALSE VAMPIRE BAT
Megaderma lyra

The Indian False Vampire Bat is a small species, only about 8 cm long. It is a woodland resident of India, Burma, Malaysia and southern China. It lives in groups of 3—50, seeking shelter during the day in caves, hollow trees, or buildings. At dusk these bats begin to hunt. They are highly predatory, catching insects, spiders, and even small birds, frogs, lizards and various species of small bats. They swoop on their prey ferociously, suck out the blood and then take small bites of flesh. They often carry their prey to the roosting site to enjoy the food in peace. After a gestation period of about 70 days, the female produces

Indian Short-nosed Fruit Bat

a single young, or occasionally twins. At this time, the females live apart from the males and keep their young with them. In India breeding usually takes

Indian Flying Fox

Indian False Vampire Bat

place in November, before the rains come. By the age of 6 weeks, the young bats are independent and accompany their mothers as they fly.

COMMON TREE SHREW
Tupaia glis

The Common Tree Shrew, which forms a connecting link between insectivores and primates, inhabits tropical Asian forests. Its length is about 40 cm, of which more than half is made up of the bushy tail. Tree shrews are active during the day and at dusk. They move nimbly along the branches, climb trunks, and leap skilfully from branch to branch. They live either singly, or in pairs, and occupy territories which they mark with a secretion exuded from a pectoral gland. At night tree shrews sleep in holes in trees, and here females give birth after a gestation period of 52—56 days. The litter size is usually two, but sometimes three or one young may be produced. They are born blind and naked, and their mother suckles them two or three times a day. The young leave the nest after 27 days to begin to forage for solid food. Tree shrews feed on insects and insect larvae, birds' eggs and the young of small vertebrates. They are fond of sweet fruit. When eating, they hold the food in their front paws and pass it to the mouth. Young tree shrews are fully grown at the age of 3 months and mature a month later.

SLENDER LORIS
Loris tardigradus

The Slender Loris is 25 cm long and has a very short tail. It moves slowly and with apparent hesitation, in the trees of the tropical forests of southern India and Sri Lanka. It ranges from low-lying areas to a height of 1 800 m in the mountains. During the day, the loris sleeps among the crowns of the trees, gripping the branches with its claws and holding its head between its legs so that it looks like a large furry ball. It wakes after sunset, stretches its limbs and begins to forage for insects and larvae in the canopy. It also takes eggs and chicks from the nests of small birds, and catches tree frogs and small lizards. It enjoys sweet juicy fruit and berries, and occasionally nibbles young shoots and leaves. This loris uses its front paws to pass the food to its mouth. Following a gestation period of 160—174 days, the female produces a single young in a tree hollow lined with soft leaves. The infant is suckled for 3—6 months, but it begins to eat soft fruit and insects when it is a month old.

SLOW LORIS
Nycticebus coucang

The Slow Loris is distributed across south-eastern Asia from Assam to Indonesia and the Philippines. This handsome loris reaches a length of 40 cm, has a very short tail and weighs 0.5—1.5 kg. Its coloration is variable. During the day the loris remains in a tree hollow and sleeps curled up in a ball. It emerges after sunset and crawls slowly along the branches in its territory, seeking prey. It feeds on insects and insect larvae, small lizards, and the eggs and young of birds. Sometimes it finds sweet fruits or nibbles bamboo shoots. After a gestation period of 185 days, the female gives birth to a single young, or exceptionally twins, in a hollow in a tree. The young loris is suckled for 3—6 months, though it often takes solid food by the age of 10 days. The natives catch these animals to keep as pets, for they are thought to bring good luck.

Common Tree Shrew

Slender Loris

WESTERN TARSIER
Tarsius bancanus

The Western Tarsier is a strange-looking animal, with large conspicuous eyes, and fingers with broadened adhesive tips. Its body is only 12—14 cm long, while its tail measures 15—20 cm and is used as a rudder when the tarsier leaps from branch to branch. The Western Tarsier is confined to the tropical forests of Sumatra, Borneo, Celebes and the adjacent islands. Although it is abundant in some localities, it is rarely seen because it is nocturnal. During the day, it hides among dense vegetation in the treetops, or occasionally makes use of a hollow in a tree. It becomes active at dusk, climbing nimbly and speedily, and often leaping distances of several metres. It grasps the branches with its broadened fingers in the same way as a tree frog, and holds its tail arched above its back. When on the ground, it can jump as far as 1.7 m and as high as 60 cm, which is an extraordinary performance for such a small animal. During its night forays, the tarsier mostly catches insects and their larvae, but it may sometimes take small geckos. At night, the male is often heard to make a twittering sound similar to the song of a bird. The tarsier is also unusual in being able to turn its head through an angle of almost 360°. After a gestation period of 185 days, the female produces a single young, which she suckles for 5—6 months.

RHESUS MONKEY
Macaca mulatta

The Rhesus Monkey is one of the best-known members of a large group of monkeys called macaques. This species is especially known for its use in medical

Western Tarsier

studies. It was as a result of experiments in blood transfusion carried out on this monkey that the Rhesus factor was discovered. The Rhesus Monkey is widely distributed from Pakistan across north and central India to southern China and south-east Asia. It was once very abundant, but its populations have been considerably reduced in some areas as a result of experimental research. Rhesus monkeys are found living on the overgrown banks of rivers, or on rocky plateaux where tall bushes provide thick cover. They live in troops of 25 or more, led by the strongest and most

Rhesus Monkey

Slow Loris

Lion-tailed Monkey

can completely ruin a harvest of bananas or oranges, so fruit growers in particular hunt them assiduously. However the Hindus in India regard these animals as sacred. At night, families of monkeys settle in small caves in the steep slopes above the rivers, or in dense, tall treetops. If they are not disturbed, they often inhabit the same sites for many years. After a gestation period of 146—180 days, females give birth usually to a single young, but sometimes to twins. These are carried on the mother's abdomen and suckled for 7—14 months. They begin to eat soft fruits at the age of 2 months. From the age of 180 days, young rhesus monkeys are capable of fending for themselves. Females mature at 3.5 years of age, and males a year later. They live to be about 30 years old.

experienced male, and to whom the others are subordinate. The leader keeps watch and warns the troop of the dangers which beset these monkeys almost continuously. Their palatable meat is a delicacy for bloodthirsty leopards; birds of prey seize the young; careless monkeys often end up in the jaws of huge pythons. On their regular daily foraging trips, rhesus monkeys nibble shoots and young leaves, seek berries and fruits, collect seeds, crack nuts or pull up juicy roots. Near human settlements they do an enormous amount of damage, raiding maize fields and taking the fruit from plantations and gardens. A large troop

LION-TAILED MONKEY
Macaca silenus

The Lion-tailed Monkey is a robust macaque, up to 125 cm long including the bushy tail, which measures 35 cm. This monkey is characterized by extremely long fur on the sides of the head, and there is a tuft at the tip of the tail, similar to that of a lion. The species is restricted to dense primary forests of south-western India, down to Cape Comorin, where it lives in small family groups of 10—20. During the day, the monkeys forage among the trees, feeding on shoots, leaves and fruit. They are also unwelcome visitors to nearby fruit plantations, causing much damage picking the

Proboscis Monkey

Siamang

unripe fruit and throwing it away. At night macaques sleep in permanent sites in tall trees hiding among the dense branches. After a gestation period of 165 days, the female gives birth to just one baby monkey, which is carried about on her abdomen. In India, the Lion-tailed Monkey is strictly protected, for it is looked upon as a sacred animal. It is, however, very rare.

PROBOSCIS MONKEY
Nasalis larvatus

The home of the Proboscis Monkey is the primary forests of Borneo. This conspicuous arboreal monkey lives near water, in the thick cover of the river banks. The male is heavily built, and is about 150 cm long including the tail, and he may weigh in excess of 20 kg. The female is smaller and less robust, reaching a weight of about 10 kg and a maximum length of 120 cm. The adult male has an enormous swollen, pendulous nose, which hangs over his mouth. It resembles a large cucumber and he has to push it away when he is eating. It acts as a resonator of sounds made by the male in order to denote his territory. Proboscis monkeys form small family groups, living in the trees. At night the families sleep in the dense treetops, beginning to forage after sunrise. They collect shoots, young leaves, berries and fruits, often descending along the lianas to the surface of the water, where they eat juicy aquatic plants. A young, inexperienced monkey may fall into the water while performing these acrobatics, but the proboscis monkeys are good swimmers and are able to get out easily, and climb up the nearest tree. After a gestation period of about 166 days, females produce a single young, weighing about 0.5 kg at birth. The mother carries her baby on her abdomen, sometimes passing it to other females, or even to its older brothers and sisters, to take good care of it. At the age of 8 weeks, the young one is able to move independently near its mother, and play with the other members of the group. Males mature when they are 6—7 years old, females by the time they are 4 years of age.

SIAMANG
Symphalangus syndactylus

The Siamang is an inhabitant of the primary mountain forests of the Mentawai Islands off the west coast of Sumatra. Here it lives at heights from 600—2 800 m. Adults stand 75—90 cm tall and weigh 8—13 kg. The Siamang has a very short tail, long fur and a large resonating throat pouch. Males often make use of this amplifying device, and their loud calls can carry over 4 km. A series of calls, made in defence of their territory, can last for as long as half an hour. Each territory covers 12—40 ha, depending on the density and the height of the trees. Siamangs live in families, each usually made up of the parents, together with 1—3, often fully grown but not fully adult young. Occasionally several juveniles may merge to form a group, and there are always some solitary monkeys looking for mates. Siamangs spend the night sitting on the branches in dense treetops. In the morning, they begin their regular foraging trips. They use their long arms to hang on branches, swing, and easily cover distances of several metres. They seek their food in the trees, feeding mainly on fruit, though they also nibble at leaves and shoots, and occasionally catch insects and larvae. Sometimes they pick termites from their nests in the trees, and sometimes they take birds' eggs or nestlings. They drink rainwater from cone-shaped leaves, or hang on branches above a river, dip one arm into the water and then suck their long, wet hairs. After 235 days of gestation, the females bear a single young, born with open eyes but only sparsely furred. At first it is carried on its mother's abdomen, only standing on its own legs after 6 months. The Siamang is mature at the age of 5—7 years.

WHITE-HANDED or LAR GIBBON
Hylobates lar

The White-handed or Lar Gibbon is distributed in south-east Asia west of the Mekong River, throughout the Malay Peninsula, south to Sumatra. It inhabits rainforests up to a height of 2 400 m. Gibbons differ widely in coloration, ranging from light cream shades through to black, but all have white-lined faces. The body length is 46—64 cm, and the weight varies from 5—8 kg. All gibbons have a stunted tail, a feature they share with the apes, and among which they were once classified, though now they form an independent group. Gibbons live in families made up of a male and a female with 1—6, often fully grown young. Solitary gibbons are either older individuals, or young animals looking for mates. Each family occupies a permanent territory of about 20 ha, which is defended by

White-handed or Lar Gibbon

Orang-utan

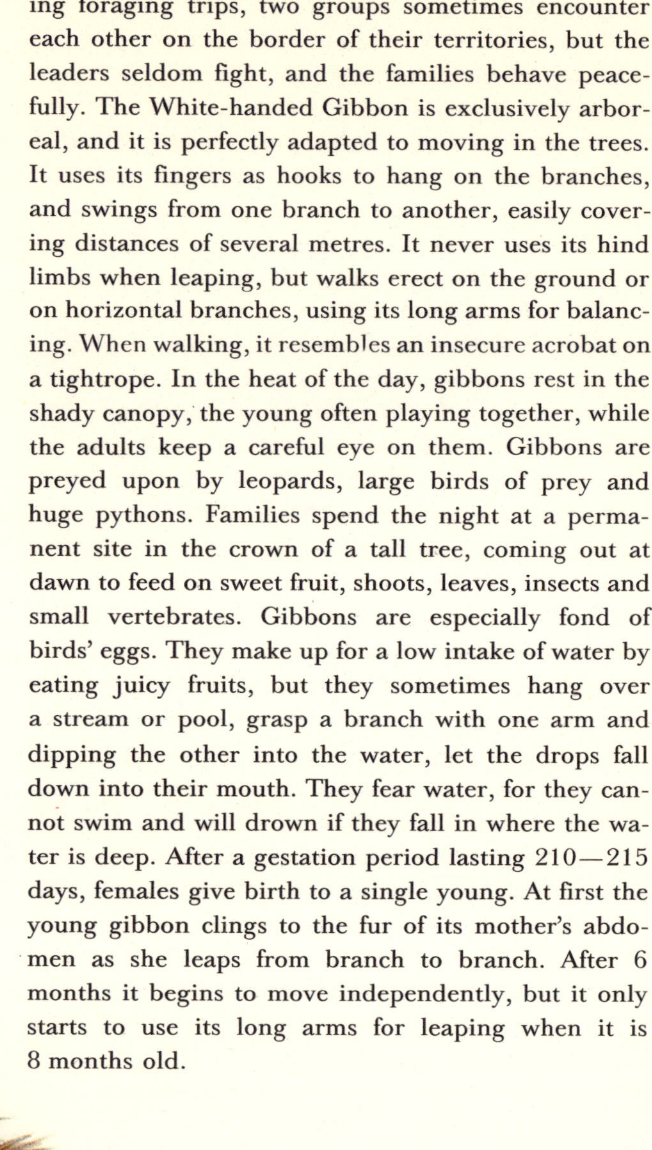

the male's far-carrying but pleasant, singing call. During foraging trips, two groups sometimes encounter each other on the border of their territories, but the leaders seldom fight, and the families behave peacefully. The White-handed Gibbon is exclusively arboreal, and it is perfectly adapted to moving in the trees. It uses its fingers as hooks to hang on the branches, and swings from one branch to another, easily covering distances of several metres. It never uses its hind limbs when leaping, but walks erect on the ground or on horizontal branches, using its long arms for balancing. When walking, it resembles an insecure acrobat on a tightrope. In the heat of the day, gibbons rest in the shady canopy, the young often playing together, while the adults keep a careful eye on them. Gibbons are preyed upon by leopards, large birds of prey and huge pythons. Families spend the night at a permanent site in the crown of a tall tree, coming out at dawn to feed on sweet fruit, shoots, leaves, insects and small vertebrates. Gibbons are especially fond of birds' eggs. They make up for a low intake of water by eating juicy fruits, but they sometimes hang over a stream or pool, grasp a branch with one arm and dipping the other into the water, let the drops fall down into their mouth. They fear water, for they cannot swim and will drown if they fall in where the water is deep. After a gestation period lasting 210—215 days, females give birth to a single young. At first the young gibbon clings to the fur of its mother's abdomen as she leaps from branch to branch. After 6 months it begins to move independently, but it only starts to use its long arms for leaping when it is 8 months old.

ORANG-UTAN
Pongo pygmaeus

The Orang-utan is found on the islands of Sumatra and Borneo, where it frequents tropical rainforests up to a height of 2 000 m. It is golden-brown, reddish-brown or fawn in colour and its fur is up to 80 cm long. The species found on Borneo in smaller, and has darker fur. A male can stand 1.5 m high, and weigh 75—100 kg, while the female is smaller, weighing only up to 50 kg. The Orang-utan is characterized by extremely long arms which in adult males can span up to 240 cm, but the hind limbs are short. Both sexes have throat pouches. These are used especially by the males as resonators when defending their territories by means of loud calls. Orang-utans live in pairs or in families containing 1—3 young of various ages. During the day, they wander through their territory, usually 30—40 m up in the canopy. They crawl slowly along the branches, sometimes hanging on them before swinging, and leaping to a neighbouring tree. They rarely descend to the ground, but when they do, they move clumsily on all fours or take only a few steps while standing erect. Both rivers and mountains represent insurmountable obstacles, effectively restricting their distribution. On foraging trips, orang-utans feed on various fruits, especially, from August to December, on the fruits of the durian tree. They also eat berries, leaves and nuts, and occasionally catch small animals, such as termites and molluscs. The families usually rest at noon, in the shade of the trees, and they settle down to sleep in the early evening. When the weather is cold or wet, they build a simple nest of branches 6—24 m above the ground, and sleep there, covered in leaves.
After a gestation period of 255—280 days, the female bears a single young, which is dependent for many months on its mother's care. It clutches her abdominal fur until the age of 3.5 months, when it begins to move on its own. Although a young Orang-utan may be suckled for 3—4 years, it begins to take solid food at the age of 6—9 months, and eats fruit and other juicy plants from one year of age. In the wild, the life expectancy of orang-utans is 30—40 years. The Orang-utan is a species threatened by imminent extinction, and is now strictly protected. It is forbidden to hunt these animals or capture them for export, for the population on the island of Borneo is estimated at 3 000, and the number of Sumatran orang-utans has dwindled to 1 000.

CHINESE PANGOLIN
Manis pentadactyla

The Chinese Pangolin is distributed from Nepal to southern China and south-east Asia, and to the islands of Hainan and Taiwan. It is a resident of tropical rainforests. The pangolin measures 100 cm including the tail, which is 50 cm long, and females are smaller than males. The pangolin has long, curved claws, which it uses for climbing. It combs the trees for termite nests, digs out the solid walls of the nest and picks out the termites with its long, sticky tongue. It catches ants in the same way. It is able to close its nostrils and ears, and so prevent attacking ants from getting inside. It hunts after sunset, sheltering by day in hollow trees, where it sleeps curled into a ball. Pangolins live in pairs or solitarily. There are 1—3 young in a litter. At birth the young are 20—30 cm long. They are able to see and are extremely active. After a few days, they follow their mother on her nocturnal trips. When settling down to sleep, the mother coils her body around the young ones.

MALAYAN PORCUPINE
Hystrix brachyura

The Malayan Porcupine lives in the tropical forests of south-east Asia, and can be seen in mountain areas up to a height of 3 000 m. It reaches a length of about 80 cm and weighs as much as 20 kg. During the day the porcupine shelters in underground burrows which it digs for itself, or in rocky clefts. It forages at dusk or early in the morning searching throughout its territory for fallen fruit, shoots and roots. It also feeds on young maize in plantations near the forests, and catches insects and other small animals. After a gestation period of 112 days, the female bears 1—4 young. These are born with open eyes and have soft quills, which begin to harden after two weeks. At this time, the young leave the burrow and wander in the surrounding countryside.

ORIENTAL GIANT SQUIRREL
Ratufa bicolor

The Oriental Giant Squirrel is found from the forests of Nepal and Burma, right through south-east Asia. This rodent dwells on hills and mountains, up to

Chinese Pangolin

a height of 2 200 m. It is about 90 cm long including the tail, which measures 45 cm or more. The squirrel is arboreal, rarely descending to the ground. It is a nimble climber, and always active, leaping from one tree to another, looking for seeds, nuts, fruits, berries, leaves and shoots. It sometimes catches insects, molluscs and young birds, and it is fond of birds' eggs. Every night, it returns to its hole in a tree to sleep, and it is here that the female bears 2—3 young after a month-long gestation period. The offspring are born blind and are not able to see until they are a month old. They leave the nest 10 days later and are weaned at 9—10 weeks of age. Oriental giant squirrels are often hunted for their beautiful fur, which is used by the natives for decorations and covers.

PREVOST'S SQUIRREL
Callosciurus prevosti

Prevost's Squirrel is indigenous to south-east Asia and Indonesia. It reaches a length of 40 cm, half of which is made up of the tail. It is predominantly arboreal, coming down to the ground only when crossing larger distances between tall trees and even then, quickly scurrying to the nearest tree trunk. It begins to forage at sunrise, moving swiftly along the branches where it nibbles shoots and blossoms, seeks nuts, fruit and seeds, preys on insects, and occasionally takes the nestlings or eggs of small birds. At night it sleeps in a hole in a tree or in a spherical nest of twigs and leaves. In the breeding season the nest is lined with fine moss and hair by the female. She bears 3—4 young after a gestation period of about 40 days. Young squirrels are born blind and are not able to see until they are 4 weeks old. Prevost's Squirrel has many enemies, especially birds of prey. The young are often taken from their nest by small tree carnivores.

COMMON GIANT FLYING SQUIRREL
Petaurista petaurista

Sri Lanka and the area ranging from eastern India to Burma are the home of the Common Giant Flying Squirrel, which inhabits densely forested areas situated at a height of 900 m. It is an extraordinary creature, equipped with a wide fold of skin which stretches along each flank between the front and hind limbs. The squirrel is over 120 cm long, but the massive tail, which is covered with long hair, takes up more than half of this length. The squirrel always stays in the treetops, usually 14—30 m above the ground. It covers distances of several metres by means of gliding jumps, stretching out the membrane like a parachute, and using its tail as a rudder. Flying squirrels can control their gliding movements, banking, and using air currents to help them, and when taking off from a height, they can leap over a distance

Malayan Porcupine

Oriental Giant Squirrel

of more than 300 m to escape their predators. The
Giant Flying Squirrel lives singly or in pairs. During
the day, it shelters in a huge nest, up to a metre
across, built in the trees and made from twigs, leaves
and moss. Sometimes the squirrel sleeps in a large
hole in a tree. It begins to forage at dusk, feeding on
leaves, shoots, seeds, fruits, berries and nuts. It also
catches insects and occasionally seizes small birds or
takes their eggs. After a gestation period averaging
40 days, the female gives birth to 2—4 blind young,
which gain their sight after 26 days. There are usually
2 litters each year.

MALAYAN BAMBOO RAT
Rhizomys pruinosus

The Malayan Bamboo Rat has short legs with very
long toes and claws adapted for digging. It attains
a length of about 40 cm, 10 cm of which is taken up
by the tail. This bamboo rat is a resident of areas of
south-east Asia and southern China, which have
a dense growth of bamboo. It can be found in mountainous areas at heights from 1 000—4 000 m.
Though it prefers to live in bamboo jungles, it is also
found among the bushes in forests. It excavates long
burrows with its feet, biting through any roots which
are in the way with its strong teeth. The Bamboo Rat
usually has several burrows, but takes up residence in
only one of them. The uninhabited burrows often con-
fuse predators such as small carnivores, causing them
to abandon their search for the rat. The Malayan
Bamboo Rat is only able to move slowly because of its
short legs. It forages after nightfall, combing the
ground for bamboo leaves, shoots, fallen fruit, seeds,
grass and fine roots. The female gives birth to 3—5
young in the underground burrow after a gestation
period of 21 days. The young are born blind, gaining
their sight after 10 days.

MALAYAN SUN BEAR
Helarctos malayanus

The Malayan Sun Bear is the smallest of the bears. It
is 140 cm long, stands 70 cm high at the shoulder and
weighs 27—65 kg. It inhabits the vast primary forests
of southern China, south-east Asia, Sumatra and
Borneo. It is solitary except in the courtship period,
when the male stays for a short time with his mate.

Prevost's Squirrel

Common Giant Flying Squirrel

Malayan Bamboo Rat

This bear is active at night, sleeping by day in a nest of broken branches or basking in the sun on an exposed branch. It is adept at climbing over rocks and up trees, using its long and powerful claws. It wanders in the neighbourhood of its den, searching the ground for fallen fruit and berries, although it climbs trees and bushes to reach them, too. It also sometimes takes unripe cobs from nearby maize fields and sweet pineapples from plantations. It searches for nests of wild bees in tree trunks, digging them out and eating the honey, and it breaks up the nests of tree termites and swallows large numbers of the inhabitants. It picks out insects and their larvae from rotten timber and occasionally preys on small vertebrates, such as frogs, lizards or rodents. Following a gestation period of 96 days, the female gives birth to two young in her well-lined den. They may be born at any time of year. The cubs are tiny and blind, and do not open their eyes until they are a month old. They leave the den after 50 days, and 10 days later they begin to forage, nibbling soft juicy fruit found on the ground. The Malayan Sun Bear reaches maturity at the age of 3.5 years.

YELLOW-THROATED MARTEN
Martes flavigula

The Yellow-throated Marten is widely distributed throughout south-east Asia, ranging to eastern Asia and the Sunda Islands. This handsome carnivore weighs up to 2 kg and is about 70 cm long including the tail, which measures 20 cm. It lives mainly in the trees, but sometimes hunts on rocky sites in the forests. It is an agile climber, running speedily along the branches and often jumping distances of up to 4 m. Hunting takes place after sunset, the marten covering over 15 km a night as it runs in the trees. It preys on squirrels and other rodents, and on small birds. It also catches insects and often eats ripe berries, and fruit from nearby plantations.

Before giving birth, the female builds a nest of moss and lichens in a hole in a tree or in a rocky cleft. After a gestation period of 220—290 days, she bears 3—5 young, which are blind at first, but gain their sight after 34 days. Young martens begin to take solid food at the age of 6 weeks.

MALAYAN STINK BADGER
Mydaus javanensis

The Greater Sunda Islands are the home of the Malayan Stink Badger. This badger measures 50 cm, weighs 1.5—3.6 kg, and has a short tail, only about 6 cm long. It is found in dry forests, where it digs out burrows for itself. As the burrow is only about 60 cm long, it is easy to see into it, although the entrance is carefully concealed under a bush or beneath tall plants. During the day, the badger sleeps in its burrow, sometimes coming out to lie in the sun. It begins to forage for food at dusk and spends the whole night combing its territory. It is omnivorous, feeding on fallen fruits, seeds, shoots, insects, frogs, lizards and snakes. It also takes fledglings or eggs found on the

Malayan Sun Bear

28

ground. In the burrow, which is lined with moss and leaves, the female gives birth usually to 3 young. They are born blind and open their eyes after 4 weeks. Young badgers are suckled for 9—10 weeks and can fend for themselves when they are 7—8 months old. They mature after 1.5 years.

BINTURONG
Arctictis binturong

The Binturong ranges from Nepal and Burma to southern China, Vietnam, the Greater Sunda Islands and Palawan. It is found only in dense primary forest, especially in the mountains, where its long hair protects it from the cold. It is about 180 cm long including the bushy, prehensile tail, which measures up to 90 cm. This sturdy arboreal civet weighs 9—14 kg. During the day, it sleeps in hollows in the trees or on thick branches. It wakes up at dusk and forages for food on its regular trails. It crawls slowly along the branches, nibbling fruits and berries and seeking birds' nests containing eggs or fledglings. It is greatly helped by its keen sense of smell.
After a gestation period of 92 days, the female gives birth to 1—2 young in a hole in a tree or in a small cave. The offspring are naked and blind at birth, obtaining their sight after 14 days. They leave the den when they are about 7 weeks old. The natives often take them from their nests and keep them as pets.

MASKED PALM CIVET
Paguma larvata

The Masked Palm Civet has a vast range of distribution. It is found in forests in southern China, the islands of Hainan and Taiwan, and from northern India across south-east Asia to Borneo. This civet is 135 cm long including its 60 cm-long tail, and it weighs 3.6—5 kg. Its coloration varies from grey to orange and yellowish-red. It has large anal glands from which it can spray its enemies with a sharp, fetid liquid, often deterring an attacker from further pursuit. These civets hide in holes in trees during the day and come out after sunset to search their territory for birds' eggs, nestlings, geckos, insects and worms. They sometimes prey on frogs on bushy river banks, and occasionally catch fish. They also like sweet fruit and roots. After a gestation period of 3 months, the fe-

Yellow-throated Marten

Malayan Stink Badger

Binturong

Masked Palm Civet

male bears 3—4 young, which reach maturity at 3 months of age.

LARGE INDIAN CIVET
Viverra zibetha

The Large Indian Civet is distributed from Assam to eastern China and south-eastern Asia. It is well-built, weighing 7—11 kg and measuring up to 130 cm including the tail, which is about 45 cm long. This civet is found in forests or among dense bushes, often near villages, where it takes chickens, ducklings and eggs in its nightly raids. It also preys on small mammals, snakes and frogs, and on fishes and freshwater crabs in shallow water. It catches its prey mostly on the ground, rarely climbing trees and never venturing very high. It also feeds on fallen fruits. During the day the civet sleeps in a hole at ground level in a tree, or in an abandoned burrow or rocky cleft. It is solitary, forming pairs only in the breeding season. A litter of 1—4 young civets is born in a hole in the ground, or among thick vegetation.

FISHING CAT
Felis viverrina

The Fishing Cat is found in Sri Lanka, southern and eastern India, Assam, south-eastern Asia, and on the islands of Sumatra and Java. It lives on the edges of forests, on river banks and the shores of lakes, and in coastal mangrove swamps. It feeds on fishes, frogs and crabs caught in shallow water, and hunts small birds, especially young ones in the reed beds. The Fishing Cat is an excellent swimmer, and is able to cross even wide streams. This handsome cat reaches a length of about 1 m including the tail, which is 30 cm long. It hunts at night and in the evening, and on overcast days it even hunts in the late afternoon. It spends the day hidden away in a hollow low in a tree, or in dense scrubland. In the courtship season, pairs spend a short time together, but the partners soon separate to live singly again. After a gestation period of 63 days, which can take place at any time of the year, the female gives birth to 1—4, though usually to 2 kittens. These are born blind and with very little fur. They open their eyes when they are 2—14 days old and venture out for the first time when they are 3—4

Large Indian Civet

weeks old. At 4—5 weeks, they begin to nibble at prey brought for them by their mother. Kittens enjoy playing in shallow water, and splashing one another. Young cats can fend for themselves at the age of 3 months, but they stay with their mother for the first year.

MARBLED CAT
Felis marmorata

The tropical rainforests which stretch from Assam and the eastern Himalayas to Burma, south-east Asia, Sumatra and Borneo are the home of the Marbled Cat. It is 130 cm long including the tail, which measures about 65 cm. This cat spends most of its life on the ground, only occasionally climbing trees to seek birds' nests. After sunset, the cat follows its regular track or lies in wait for its prey, pouncing on small mammals, especially rodents, and sometimes catching small lizards. During the courtship season the cats form pairs, but afterwards the partners return to their respective territories. The female gives birth to 1—4 young in a hole near the bottom of a tree after a gestation period of 65 days. The kittens are born blind, but are able to see after a week.

GOLDEN CAT
Profelis temmincki

The Golden Cat is a large carnivore measuring up to 160 cm including the tail. It is distributed from Nepal to southern China, and across the whole of south-east Asia to Sumatra. Although it is abundant in some lo-calities, it is rarely seen because of its nocturnal way of life. During the day this cat shelters in holes in trees or in caves in primary forests, settles in abandoned burrows or sleeps among the bushes. At dusk it comes out and follows a well-beaten track through its territory. It usually moves on the ground, but it is a good climber and often preys on nestlings in the trees. It usually hunts only small mammals, but it sometimes attacks the young of small species of deer, such as the Muntjac. For most of the year, these cats live singly, but in the breeding season the male seeks a mate and stays with her for a few days. After a gestation period of about 66 days, the female bears 1—5 young in a sheltered burrow. The kittens are born blind and obtain their sight after about a week.

CLOUDED LEOPARD
Neofelis nebulosa

The Clouded Leopard is undoubtedly one of the most attractive of the Big Cats. It is found from eastern India and Nepal to southern China and Taiwan, and across south-east Asia to Sumatra and Borneo. It reaches a length of 160—190 cm including the tail, which measures 70—90 cm. The leopards of Taiwan have rather shorter tails. The Clouded Leopard is a denizen of tropical forests. It is solitary for most of the year, forming pairs only for a short time in the mating season. After a gestation period of 86—92 days, the female bears 1—5 young in a hollow tree or in a rocky cave. Born blind, the young leopards are able to see after 10—11 days. At the age of 35 days, they leave their shelter for the first time, and play in the sun in front of the lair. When they are 40—45

Fishing Cat

Marbled Cat

Golden Cat

days old, they begin to nibble at the flesh of prey which has been brought to them, but they continue to be suckled for 5—6 months. Males mature after 3 years, females after 2.5 years. Leopards hunt during the day and at dusk, but in places where they are hunted themselves, they prefer to prowl at night. They climb branches, preying on monkeys and other mammals in the trees, while on the ground they pounce on young deer and catch birds. When food is in short supply they live on small rodents and lizards. The Clouded Leopard has been hunted to extinction in many areas for its beautiful and valuable fur, so it is now protected by law.

BENGAL TIGER
Panthera tigris tigris

India and the northern region of south-east Asia are inhabited by the Bengal Tiger. This massive member of the Big Cats has a total length, including the tail, in excess of 350 cm, and its height at the shoulder may reach 100 cm. The Bengal Tiger lives in jungle forests, in bamboo thickets, and in extensive reed beds, while in the south it can be encountered in the mountains, up to a height of 3 000 m. Each tiger occupies a vast territory and within it, takes up residence in a permanent lair in a cave beneath a large fallen tree, or in a sheltered bush. The tiger hunts mostly by day, but sometimes it prowls at dusk, usually following the same beaten tracks. It crawls unseen through the thick cover, its crosswise stripes blending perfectly with the leaves of tall bamboo and reed stems as it stalks wild hogs, deer or antelope. The tiger makes a few formidable jumps, then pounces on its prey. Its jaws crush the animal's neck or slash the throat of bigger mammals. If the intended victim takes flight, the tiger rarely gives chase, preferring to lie in ambush once more. It drags its prey to a spot near water, where it can both eat and drink. An adult tiger usually consumes 7—9 kg of meat at a time, but a hungry one can devour more than 40 kg. When the tiger is satiated, it covers the remains of the kill with leaves and returns to feed on it later. The tiger hunts animals of all kinds, including fishes, catching them skilfully in the shallows. Also it occasionally consumes fallen fruit. Tigers only seldom become man-eating, and those which do have to be sought out and killed. For most of the year, tigers are solitary. In the mating season the male joins his mate and lives with her for a few weeks. When two males are interested in the same female, they fight fiercely. When meeting his chosen mate, the male greets her in a low, rumbling voice, and the female answers. After a gestation period of 95—114 days, at any time of year, the female gives

Clouded Leopard

Bengal Tiger

birth to a litter of 1—6 young, although usually she has twins or triplets. The cubs are born blind and gain their sight after 2—7 days. They are suckled for 5—6 months, after which time they follow their mother on hunting trips and feed on the prey which she catches. They begin to fend for themselves at the age of 11—15 months, but stay with their mother for as long as 4 years. For this reason, a tigress in the wild has only one litter every 3—4 years. Young females mature after 2.5 years. The males mature one year later, and become fully grown at the age of 5—6 years. The Bengal Tiger is a good swimmer, often crossing large rivers and straits. It is, however, a very bad climber, and prefers to move on the ground, where it can cover distances of up to 10 m by jumping.

INDIAN ELEPHANT
Elephas maximus

The Indian Elephant has always been one of the best-known animals. It was used for labour more than 5 500 years ago in India, and even today elephants are used to carry heavy burdens, to fell trees, or pull up smaller trees by the roots. Both young and adult elephants are caught in the wild and are then trained, which is why, over thousands of years, elephants have been tamed but never domesticated. The Indian Elephant is the second largest of the terrestrial animals, only the African Elephant being larger. It reaches a height of 310 cm at the shoulder and a weight of over 5 tonnes. Females do not usually have tusks, though older animals sometimes have rudimentary ones. It is not unusual for males to have only rudimentary tusks; indeed in Sri Lanka, tusks are present in only 11 per cent of males. They measure 160 cm at most, and weigh 20 kg.

The Indian Elephant is indigenous to south and south-east Asia, where 4 subspecies are distinguished. The subspecies *Elephas maximus bengalensis,* which inhabits south-western, southern, eastern and north-eastern India, Burma and south-eastern Asia, is the most common. The Ceylon Elephant *(Elephas maximus maximus)* is found in Sri Lanka, and the rare Sumatran Elephant *(Elephas maximus sumatranus)* lives in Sumatra. The Malayan Elephant *(Elephas maximus hirsutus)* was probably introduced to Malaysia and Borneo many years ago, but it has been almost exterminated.

Elephants are sociable, living in communities of 15—30 individuals made up either of males, or of females and their offspring. Old males sometimes live singly, although there is some contact between them. In the rutting season, adult males often fight each other near the herds of females.

Elephants roam the primary forests up to a height of 3 000 m, ascending the Himalayan slopes almost to the snowline. They favour large bamboo thickets, swamps and marshland, seeking wet situations in times of drought, and moving to drier woodland in the rainy season. Groups of elephants follow regular paths leading to water. These paths branch out into

Indian Elephant

scores of side paths, which lead the animals to their sources of food. Elephants are mainly active during the day and in the evening, though sometimes they feed on clear nights. They sleep for only 2—4 hours. On their forays, they travel at a speed of about 7 km per hour, sometimes doubling this by trotting. Elephants spend most of the day feeding. They eat grass, leaves, twigs and various fruits, or even soft wood, which they crush with their four huge molars. Herds of elephants sometimes wander into banana plantations or cultivated fields, causing considerable damage to the crops. Elephants have sensitive skin, so they spend several hours a day bathing and wallowing in mud or throwing sand or earth all over their bodies. After a gestation period averaging 21.5 months (in India and Burma it lasts 17.5—25 months), a single young is born. Twins are exceptional. The young elephant, weighing 50—120 kg, stands up a few hours after birth and follows its mother and the rest of the herd from the second day. It suckles directly with its mouth for 8—10 months. From the age of 4 months, it begins to pluck soft grass or leaves, passing the food into its mouth with its trunk in the same way as the

adults. After 6 months, its diet of milk is regularly supplemented with shoots, fruits and leaves.

BABIRUSA
Babyrousa babyrussa

The Babirusa is a species of wild pig, which reaches a weight of 90 kg and a length of about 100 cm. Its tail is about 30 cm long. The Babirusa is confined to the island of Sulawesi and small adjacent islands, but in many areas, it has been hunted to extinction, mostly by the native population, for its palatable meat. Babirusas frequent moist localities in forests, or settle in reed beds along the banks of rivers. They can also be encountered in thick coastal vegetation. They hide away in thickets during the day, coming out to forage at dusk or at night. Some males live together with the females and their offspring throughout the year, but more usually they are solitary, seeking mates only in the breeding season, which is generally in October. After a gestation period of 137—155 days the female gives birth to twins, which she suckles for 2—3 months. After 2—3 weeks the young leave their shelter, and begin to follow their mother. These pigs feed mainly on leaves, shoots, fruits and water plants — for babirusas are efficient swimmers. They also dig for larvae in soft soil or rotten wood.

LESSER MALAY MOUSE DEER
Tragulus javanicus

The dense, high rainforests of south and south-east Asia, and the Greater Sunda Islands are inhabited by the Lesser Malay Mouse Deer. It stands only 20—30 cm high at the shoulders, is about 65 cm long, and weighs 2.5—4.5 kg. These ungulates live in pairs or in families, but older specimens are solitary. The mouse deer hides away during the day, in the abandoned burrows of other animals, in rocky clefts, or in thickets. In the evening it comes out to feed on grass, shoots, leaves, fallen fruits and seeds. It also eats insects and occasionally small rodents. The mouse deer seeks its food mainly in thickets, where it darts about in the undergrowth. When escaping an enemy, it often seeks refuge in water, for it is an excellent swimmer. The female has one, or sometimes two young, following a gestation period of 150—155 days.

INDIAN MUNTJAC
Muntiacus muntjak

The Indian Muntjac, a small species of deer, is widespread throughout Sri Lanka, southern and south-east Asia, the Greater Sunda Islands and the island of Bali. It is about 50 cm tall, 100 cm long and weighs 14—18 kg. The male has small antlers, up to 15 cm long, with a single tine. The Indian Muntjac is found only where there is thick undergrowth, in primary forests, on river banks, and sometimes in city parks where there is dense cover. It lives both in low-lying areas, and on mountains up to a height of 4 000 m. It spends the day, singly, in pairs or in small family groups, asleep in a thicket. Individuals have permanent territories with regular paths, which they mark with secretions exuded from a gland on the chin. Muntjacs begin to forage at dusk and remain active all night, feeding on leaves, shoots, twigs, grass and various fruits, and grazing when they can on cultivated crops.

Following a gestation period of about 185 days, the fe-

Babirusa

male gives birth to one, or sometimes two young, which are suckled for 4—5 months and which mature after 6 months. Indian Muntjacs are preyed upon by tigers and leopards.

SAMBAR
Cervus unicolor

The Sambar is found throughout southern and south-east Asia, south-western China and Taiwan, Sumatra and Borneo, and it has been successfully introduced in Australia, New Zealand and Florida. The Sambar stands 120—155 cm high at the shoulder, and has characteristic glandular projections on its eye sockets. The male grows massive, backward-curved antlers

Lesser Malay Mouse Deer

Indian Muntjac

which have only a few tines. For most of the year sambars live singly, although young males sometimes form small groups. In the rutting season, several females may gather to form a small herd. After a gestation period of 245—284 days, a single young is born in a sheltered place, in any season of the year. After a few days the calf follows its mother, but it continues to be suckled for 7—8 months. Sambars are normally active during the day, although they may graze at dusk as well. However, in places where they are hunted, they venture out only at night. They feed on grass, shoots, twigs, fruits, leaves, reeds, lichens and moss. After sunset, they sometimes visit rice paddies and nibble the seedlings there. They enjoy bathing and swim well, often staying in the water for several hours. Sambars fall prey to tigers and leopards, and the young are often seized by other carnivores.

AXIS DEER or CHITAL
Axis axis

The Axis Deer, or Chital, is one of the most attractively coloured species of deer, the upper parts being marked during part of the year, with small white

Sambar

spots. It is native to Sri Lanka and India, where it lives in large herds in thinly forested areas, usually near water. This deer takes refuge in the water when it is in danger, for it can easily swim across large streams. Outside the rutting season, the males sometimes gather in groups. The Axis Deer is herbivorous. It grazes predominantly during the day, but in places where it is pursued by its enemies, it feeds at night. It is about 80 cm tall, 105—175 cm long, and weighs 27—45 kg. The males grow thin antlers with several tines, which are usually shed in August. After a gestation period of 210—230 days, the female gives birth to 1—2, or sometimes even to 3 young, mostly in April or early May. They are suckled for 4—5 months and mature after 1.5—2 years. The Axis Deer is often bred in captivity and it has been successfully introduced in the wild in New Zealand, Australia and Hawaii.

BANTENG
Bos javanicus

The Banteng is one of the most handsome of the wild oxen. Unfortunately, it is also one of the rarest of the mammals. Its range of distribution covers Java, Borneo, eastern Sumatra, Bali and the Malay Peninsula. There are only some 400 specimens left on Java, where it is relatively abundant, but all the other areas put together contain less than this. The Banteng is now strictly protected, but although the governments of the countries in which it is found prohibit the hunting of this animal, the natives ignore the laws and continue to hunt it. The Dayak tribesmen of Kalimantan and Sarawak prize banteng heads, displaying their trophies as once they did the skulls of their human enemies. The number of bantengs has also fallen because their original habitats have been destroyed by logging. Bantengs live in forests, especially bamboo jungles, at heights of about 2 000 m. They often feed on grass, leaves and the shoots of bushes, but they prefer young bamboo shoots. Bantengs live in small groups of ten or so. The herds are composed of cows and calves, while old bulls live singly, and young males often form separate groups. When the herd is grazing or resting in the shade, one older cow always stands guard, warning the others of approaching danger by stamping on the ground. After a gestation period of 270 days, at any time of year, the female bears a single young, which is often able to

Axis Deer or Chital

stand up after half an hour and which can suck within an hour. It is suckled for 6 months, but from 2 months of age the calf is able to nibble grass and leaves. It eats solid food regularly from the age of 4 months. Adult males are dark brown, while females are buff in colour. The Banteng has been domesticated on the islands of Bali and Java.

NILGAI or BLUE BULL
Boselaphus tragocamelus

The Nilgai, or Blue Bull, is native to the open plains of India. The male stands about 150 cm high at the shoulder, weighs approximately 200 kg, and has small, curved horns. The female is smaller, brownish in colour and lacks horns. The bulls acquire their darker coloration after three years. The Nilgai lives in small groups of 6—20 cows with their young, while adult bulls are solitary, and only join the herds in the rutting season. After a gestation period of 240—260 days, the female usually produces twins, though sometimes she has a single calf. She suckles her offspring for 4—6 months, but from the age of 2 months, the calves are able to nibble grass, shoots, leaves and fruits like adult antelopes. The Nilgai traverses regular, well-trodden paths, always using the same areas

Banteng

for rest and grazing. It is heavily preyed upon, especially by tigers and leopards.

BLACK BAZA or BLACK-CRESTED LIZARD HAWK
Aviceda leuphotes

The Black Baza, or Black-crested Lizard Hawk, is widespread throughout southern and south-east Asia, being found in primary forests in both lowlands and mountain localities up to a height of 1 500 m. This bird of prey is about 35 cm long and it has a conspicuous, tall crest. For most of the year it lives singly or in family groups, as many as 20 birds hunting together. Bazas catch large insects, such as the biggest cicadas, both on the wing and on the trees, and they prey on small lizards and frogs. At dusk, they sometimes pursue bats. In the breeding season, pairs form and each pair builds a nest of twigs in a tall tree. The nest has a lining of grass and green leaves, which is replaced from time to time. The female lays 2—3 greenish-white eggs, sometimes patterned with a few russet spots. The partners take turns at sitting on the clutch, but the female incubates more frequently, spending almost 33 days on the nest. The male makes hooting sounds, twice or three times repeated.

CRESTED SERPENT EAGLE
Spilornis cheela

The Crested Serpent Eagle is native to south and south-east Asia, the Greater Sunda Islands and Palawan, and is abundant in all these areas. It lives in dry woodland on the edges of rivers and lakes, and can be found in the Himalayas, up to a height of 2 000 m. This beautiful raptor is about 70 cm tall. It is often seen in tea plantations, where it perches on tall trees and scans the ground for small snakes, both venomous and non-venomous. It also catches lizards, frogs, small rodents, birds, freshwater crabs and occasionally domestic fowls.
A pair usually stays together for life, but outside the breeding season, solitary birds can sometimes be seen hunting in their vast territories. The nesting season

varies according to the area of distribution, being from February to May in Sri Lanka. Both partners help to build a medium-sized nest in a tall tree. It is made of twigs and is lined with green leaves. The female lays a single egg, or sometimes two. The eggs are white, covered with russet spots. The female incubates them for 35 days, while her mate brings food to her. Young eagles are fed mainly on slender snakes of the genus *Ahaetulla*. Serpent eagles have very keen sight, and are able to detect these snakes despite their protective green coloration.

BESRA SPARROWHAWK
Accipiter virgatus

The Besra Sparrowhawk inhabits the forests of south and south-east Asia, the Greater Sunda Islands and the Philippines. This raptor stands up to 35 cm tall, the male being slightly smaller than the female and differing in coloration. The female has chocolate brown plumage on her back, while the nape and the crown of her head are black. The species can be seen in abundance on wooded hills and in mountains up to a height of 2 000 m, but it seldom occurs in low-lying situations. During the nesting season, the birds establish themselves in dense jungles, but after the young have fledged, they seek open countryside. Sparrowhawks normally build nests of twigs in the crowns of

tall trees, but they sometimes use nests abandoned by other birds, and just line the nesting cup. The female lays 3—5 eggs, which vary in colour, though they are usually blue, with russet and brownish-black spots. The female incubates the clutch for about 36 days, and though the male may sometimes relieve her, for

Nilgai or Blue Bull

*Black Baza
or Black-crested
Lizard Hawk*

Crested Serpent Eagle

39

Besra Sparrowhawk

the most part he occupies himself only with hunting, passing food to his mate for her to divide it into portions for the young. In due course the female begins to hunt as well. Sparrowhawks catch squirrels, bats, small birds such as barbets or tits, lizards and large insects.

GREY-FACED BUZZARD-EAGLE
Butastur indicus

The Grey-faced Buzzard-Eagle reaches a height of about 42 cm. During the nesting season, in spring and summer, it is found in forested areas in the southern Japanese islands, Korea and adjacent parts of China and Russia. The nest is built 9—12 m above the ground on the branches of large trees. It is made of moss, twigs, grass stalks and leaves. The female lays 3 eggs, usually greenish-white, and sometimes with pale red spots. She incubates them alone for 29 days. The young are fed by both parents. On fledging, the buzzards migrate south to spend the winter in the forests of south-east Asia, the Greater Sunda Islands, and even the Philippines. They have been known to fly as far as New Guinea. The diet of the Grey-faced Buzzard-Eagle consists of snakes, lizards, small mammals and insects, particularly grasshoppers.

PHILIPPINE MONKEY-EATING EAGLE
Pithecophaga jefferyi

The Philippine Monkey-eating Eagle is an enormous raptor standing up to 100 cm high and weighing 4.5 kg. With its massive curved beak and powerful sharp claws, it is the scourge of monkeys living in the primary forests of the Philippines. Whenever its silhouette passes high above the trees, the macaque monkeys shriek with fear and try to hide out of sight of their fearsome enemy. The eagle glides gracefully overhead, apparently indifferent to the panic below, but suddenly, like a guided missile, it swoops down onto the target it has pinpointed. It follows its victim among the branches, catches up with it in an instant, and pierces it through with all eight talons. In addition to monkeys, this eagle hunts flying lemurs of the genus *Cynocephalus* and squirrels of the genus *Callosciurus*. It sometimes takes domestic poultry from the

Grey-faced Buzzard-Eagle

Philippine Monkey-eating Eagle

outskirts of villages, and as a result is itself hunted by the local people.

The Philippine Monkey-eating Eagle lives in mountain forests at heights from 200—1 300 m, nesting on branches of the tree *Sapium luzonicum,* 30 m above the ground. The nest is made of twigs and is lined with green leaves. During the first 3 weeks of incubation, both birds, though mainly the female, keep adding more lining to the nest. The female usually lays a single white egg, which she incubates for 55 days. The male brings her food, and later he feeds the young eaglet as well, passing prey to his mate near the eyrie. A young monkey-eating eagle takes its first flight at the age of 13 weeks.

CRESTED HAWK EAGLE
Spizaetus cirrhatus

The Crested Hawk Eagle is distributed throughout south and south-east Asia, frequenting forested localities up to a height of 2 300 m. It reaches a length of about 70 cm, the male being the smaller of the two sexes. This handsome raptor also occurs in a black form. It prefers the edges of primary forests and clearings, where it perches on dead branches and basks in

the sun with its wings wide open. Here it also waits for prey, such as lizards, birds and small rodents. This raptor often visits villages and takes chickens, and a pair of hawk eagles living near a farm can cause a great deal of damage. On sunny days hawk eagles bathe in shallow streams and lakes. Each pair builds a large nest 12—30 m above the ground in a tall tree, especially in trees of the genera *Ficus* and *Albizzia.* The nest is constructed from branches and measures 1 m across and 45 cm high. The birds line the nesting cup with fresh green leaves. The female lays a single egg, usually white, though sometimes it is patterned with a few russet spots. She incubates the egg for about 40 days, and the male brings food to her and later to the whole family.

RUFOUS-THIGHED FALCONET
Microhierax caerulescens

The Himalayas and south-eastern Asia are the home of a small bird of prey, the Rufous-thighed Falconet, which stands only 18 cm high. It occurs on the edges of forests up to a height of 2 000 m, but in the hills it is usually encountered at about 600 m above sea level. It lives in pairs, but after the young have fledged, the

Crested Hawk Eagle

birds stay in families for some time. Pairs build their nest in a hole in a tree abandoned by woodpeckers or barbets. The female usually lays 4 whitish eggs. Although this raptor is quite common, little is known of its nesting habits or how it cares for its offspring. The Rufous-thighed Falconet feeds mainly on insects, particularly grasshoppers and dragonflies, caught on the wing. It also preys on large butterflies as they suck

Rufous-thighed Falconet

nectar from flowers, and it may occasionally catch a small bird.

RING-NECKED PHEASANT
Phasianus colchicus

The Ring-necked Pheasant is one of the most widespread pheasant species. Its range of distribution stretches from the Caucasus to eastern and south-eastern Asia, but distribution is not continuous. Some subspecies occur in very small areas, sometimes hundreds of kilometres from the place where another subspecies is found. This phenomenon is particularly apparent in China, where the formation of deserts resulted in the pheasants being isolated in pockets of forested land. As a result independent subspecies differing in size and coloration were created. Within the vast woodland areas the subspecies sometimes hybridize, giving rise to forms often described as new subspecies. Altogether there are about 33 subspecies of the Ring-necked Pheasant.

The cock is 75—90 cm tall and has a bald, bright red area around his eye. The hen is only about 60 cm long. Her plumage is spotted grey and brown and her brown tail has dark crossways stripes. The Ringnecked Pheasant is found in thin woodland at heights from 600 to 3 300 m. It is predominantly a mountain species, never leaving its native territory. It is not a very good flier, being capable of rapid but not sustained flight, so after a steep take-off, it soon glides down again. The male defends his own territory, sometimes engaging in fierce fights with rival males. He is polygamous, acquiring a group of 3—5 hens with which to mate. During the courtship period in spring, the cocks make high-pitched sounds as they strut, and whirr their wings. Then they run around the hen, with their wings hanging low and their heads bowed. After the nuptial period, the cock takes no further interest in the hen or her family. The female makes a scrape in the ground with her body, usually in a clump of grass or at the foot of a tree, lining it with dry leaves and grass. She lays 8—15 eggs, whitish to olive-green in colour and sits on the nest for 22—27 days. The young hatch early in the morning and leave the nest in the afternoon, guided and protected by their mother. They have no permanent roosting place, and while they are incapable of flight, they hide in tall grass or among bushes, and she covers them with her wings at night. When they are

Ring-necked Pheasant

half-fledged, the chicks roost with her in trees. Adult pheasants usually spend the night among the branches, but in summer they sometimes sleep on the ground. Pheasants forage in the morning and in late afternoon, spending the heat of the day in the shelter of the undergrowth. Young pheasants consume vast numbers of insects, spiders and molluscs. They also feed on small leaves and grass. At the age of 14 days, they begin to flutter about near the ground, and they are able to fly properly when they are 45 days old. They stay with their mother for about 2 months and then begin to fend for themselves. Pheasants are ground feeders, seeking mainly seeds, berries and green plants, though they sometimes catch small invertebrates, and the occasional young lizard, or snake. Pheasants drink regularly, usually between 10 and 11 o'clock in the morning.

CEYLON SPURFOWL
Galloperdix bicalcarata

The Ceylon Spurfowl is native to Sri Lanka. The male reaches a length of 34 cm and has two characteristic spurs on each tarsus. The upper spur is 15—30 mm long, and the lower measures 10—20 mm. Sometimes the female also has spurs, but they are smaller. The hen is dark brown and has a white throat. The Ceylon Spurfowl inhabits wooded hills and mountain slopes up to a height of 2 000 m, being most commonly found in the rainforests of the southern part of the is-

land. It lives a secluded life in the shelter of the thickets, from where its voice is often heard. It lives in pairs. The female builds her nest on the ground beneath a bush or in tall grass, and lines it with leaves and stalks. She usually lays only 2 eggs, but may lay as many as five. The eggs are a uniform cream or pinkish colour. The hen sits on the clutch for 3 weeks. The diet of this spurfowl consists of termites, various other insects, seeds and green plant material, all found on the ground.

RAIN QUAIL
Coturnix coromandelica

The Rain Quail is distributed through Sri Lanka and India to Burma. It is about 18 cm long. The hen has

Ceylon Spurfowl

Rain Quail feeds on small seeds, green plant material and small invertebrates.

KOKLAS PHEASANT
Pucrasia macrolopha

The Koklas Pheasant is common in the Himalayas from Afghanistan to central Nepal, and it also occurs in north-eastern Tibet, extending to eastern China, its distribution being discontinuous. The male measures 58—64 cm. The female measures only about 55 cm, and is spotted black and brown above, with fine buffish lines. The crown of her head is a pale chestnut colour and she has a short crest. Her throat is creamy-white, and her underparts are pale brown with black spots. The Koklas Pheasant inhabits slopes at heights from 1 500—4 500 m, sometimes ascending to 5 000 m in summer, and descending to valleys and localities at an altitude of 1 200 m in winter. It lives predominantly in coniferous forests, and seeks the heavy undergrowth of bamboo jungles. It is usually solitary except in the nesting season when it forms pairs to breed, but some pairs stay together throughout the year. In winter, the birds often stay close together, but they never form flocks. The breeding season takes place from April to June. The hen lays 5—7, or sometimes even 9 eggs, in a shallow depression under a bush. The eggs are buff in colour, with blackish-brown spots. The female incubates the clutch for 20—21 days. At the age of 5—7 days, the young pheasants begin to fly. This species feeds on seeds, shoots, green plants, insects and various other small invertebrates, found on the ground.

a brownish ground colour covered with grey spots, and a pale beige pattern on her head. This quail normally dwells in hilly situations, but can also be found in the mountains up to a height of 2 000 m. It lives always in bush-covered localities or along the edges of jungles. For most of the year it lives singly or in pairs, and sometimes it stays in family groups of 5—8 birds. It is monogamous and forms pairs in the breeding season. The female builds her nest on the ground in a clump of grass, lining it with grass stalks. She lays 4—10 yellowish eggs, densely marked with dark, blurred patches and spots, and incubates the clutch by herself for 18—19 days. The young are protected by both their parents. Nesting usually takes place between March and September, and when the brood is fledged, the families of northern India merge to form flocks and migrate to Sri Lanka for the winter. The

RED JUNGLE FOWL
Gallus gallus

The Red Jungle Fowl is distributed in five subspecies in southern and south-eastern Asia, and it also occurs in southern Sumatra, Java and Bali. The cock reaches a length of 70 cm, but the hen is only 40 cm long. She is predominantly greyish-brown, with lighter underparts and tawny-edged neck feathers. The Red Jungle Fowl frequents vast forests, from low-lying situations up to a height of 1 500 m. For most of the year it lives in flocks, breaking up into groups of one cock and 5—6 hens in the breeding season. The hen lays 5—7 eggs, off-white in colour, in a nest on the ground. She incubates them for 21 days and rears the brood by herself. As soon as they are able to fly the chicks begin to roost in the trees with their mother. This species feeds on the ground, seeking seeds, berries, green shoots, insects and their larvae, various invertebrates and the young of small vertebrates, such as geckos and agamas.

The Red Jungle Fowl is the ancestor of domestic fowls. It is known that fowls were domesticated as early as 3 200 B.C. in India, for they are depicted in statues and paintings on pottery, representing the Indian culture of that time. Over the centuries the single species developed into many domestic breeds, including the Game Cock, males of which were once used (and still may be in some places) for cock-fighting. This form of entertainment, probably originating in religious rituals, appeared in south-east Asia and on the Indonesian islands as early as 3 000 B.C. An outstanding ornamental variety is the Japanese Phoenix, produced as a result of selective breeding on the island of Shikoku. The tail feathers of this breed are often more than 3 m long, and the cocks have to roost on very high perches if they are not to damage them. Furthermore, when a cock alights on the ground, his owner watches anxiously over him and carries his train behind him.

LA FAYETTE'S JUNGLE FOWL
Gallus lafayettii

La Fayette's Jungle Fowl is native to Sri Lanka. The cock reaches a length of 72 cm, while the hen is only about 36 cm long. She has a predominantly russet-brown mantle and black spots on her wings. Her breast and flanks are russet-brown with beige speckles, and the rest of her body is black, covered with white spots. La Fayette's Jungle Fowl dwells in mountain forests, usually at heights from 1 600 to 2 000 m. Breeding takes place throughout the year, but mostly from December to April, each cock mating with several hens. The female builds her nest on the ground, lining it with dry leaves, roots and stalks. She may occasionally nest in a tree, using a nest abandoned by raptors, 3—4 m above the ground. She normally lays 3—5 eggs, pale cream or pink in colour, and often delicately marked with fawn spots. She incubates the clutch by herself for 20—21 days. The diet of this bird consists of plant material and small invertebrates.

Red Jungle Fowl

HIMALAYAN FIREBACK or KALIJ PHEASANT
Lophura leucomelana

The Himalayan Fireback, or Kalij Pheasant, is found in 9 subspecies, from the western Himalayas to northern Malaysia, where it inhabits mountain forests at heights from 400 to 3 800 m. The males measure up to 70 cm in length, and the various subspecies differ in coloration. The hens are smaller, and are usually buff in colour, with pale, scaly spots. The hen usually

La Fayette's Jungle Fowl

*Himalayan Fireback
or Kalij Pheasant*

lays 6—9 eggs, though larger clutches are sometimes produced. The eggs are yellowish, pale cream or pink in colour. The nest is made on the ground beneath a bush, and is lined with leaves. Here the hen incubates her eggs for 24—25 days. When the brood is fledged, the firebacks stay together in coveys of 10 to 20. They feed on small invertebrates and on plant material found on the ground.

SWINHOE'S PHEASANT
Lophura swinhoei

Swinhoe's Pheasant is indigenous to Taiwan, where it was discovered in 1862 by the English naturalist Swinhoe. At present, it is one of the rarest bird species in the wild, although it is relatively common in captivity. The male is up to 80 cm long including the tail, which measures 50 cm. Cocks moult into their adult plumage when they are 2 years old. The female measures about 50 cm, and has a chestnut brown ground colour covered with small pale black-rimmed

spots. Her central tail feathers are brown with light and dark bands. She lays a clutch of 6—12 reddish to cream-coloured eggs and incubates them for 25 days. This species feeds on plant material and small invertebrates, found on the ground.

CRESTED or MALAYSIAN FIREBACK
Lophura ignita

The Crested or Malaysian Fireback is a resident of the Malay Peninsula, Sumatra and southern Borneo. This pheasant has a characteristic tall crest. The male reaches a length of 70 cm, and the 4 subspecies differ in coloration. The hen is about 56 cm long, and also has a tall crest. She has russet upper parts, wings covered with fine black spots, black tail feathers, and a white chin and throat. Her breast and flanks are chestnut coloured or black and brown with white-tipped feathers. This species usually lives in pairs, and inhabits dense primary forest and bamboo thickets. The hen lays 4—8 cream-coloured eggs in a leaf-lined nest on the ground, and incubates the clutch by herself for 24 days. This pheasant feeds on seeds, green plants and small invertebrates.

SIAMESE FIREBACK or FIREBACK PHEASANT
Lophura diardi

The Siamese Fireback, or Fireback Pheasant, is native to Burma, Thailand and southern Vietnam. The cock reaches a length of 80 cm. Young males moult into their adult plumage during the first year, but they are darker in colour than more mature males, and have shorter tail feathers. The hen is about 60 cm long, and

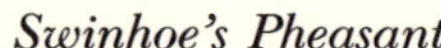

Swinhoe's Pheasant

Crested or Malaysian Fireback

has a dark brown head, throat and neck. Her upper mantle and underparts are buff in colour, while her wings, lower mantle and tail are fawn with black stripes. The feathers on her flanks have white edges, and the two pairs of middle tail feathers are black with broad brown and white stripes. The Siamese Fireback is found in primary forests and in bamboo jungles up to a height of 700 m. During the day it often basks in a warm, sunny place, but there is always a thicket nearby to give shelter in case of danger. The female lays 5—8 eggs, varying in colour from pinkish to dark tawny, in a nest on the ground. She sits on the nest by herself for 24—25 days. This pheasant feeds on insects and on a variety of small seeds, found on the ground.

ELLIOT'S PHEASANT
Syrmaticus ellioti

Elliot's Pheasant is native to south-eastern China, where it frequents open woodland at a height of 1 400 m. The male reaches a length of 80 cm including the tail, which measures about 40 cm. The female is about 50 cm long and mainly russet in colour. Her chin and throat are black, her mantle is covered with tiny whitish spots, and her breast feathers are brown with white edges. For most of the year this pheasant moves in small coveys. During the nesting season it usually forms pairs, though sometimes the male acquires two hens. In early March the female lays a clutch of 6—8 cream-coloured or pinkish eggs, and incubates them for 25 days. Young males are soon

easily recognized by their striped central tail feathers. These are a chestnut colour in the females. Elliot's Pheasant is a ground feeder, seeking seeds, berries, green shoots and small invertebrates.

MIKADO PHEASANT
Syrmaticus mikado

The Mikado Pheasant inhabits the centre of Taiwan, where it frequents mountain forests at heights from 2 000 to 3 300 m. The cock reaches a total length of 90 cm including the tail feathers, which measure

Siamese Fireback or Fireback Pheasant

Mikado Pheasant

Elliot's Pheasant

Reeve's Pheasant

REEVE'S PHEASANT
Syrmaticus reevesii

Reeve's Pheasant is common in northern and central China. The cock can reach a length of 210 cm including the long central tail feathers, which measure about 180 cm. The female is only 75 cm long. She is dark greyish-brown above, with black-edged feathers, and predominantly ochre below, with rust brown spots. Her tail is grey, with black crosswise stripes, and measures up to 45 cm in length. This species is found in forests at heights from 300 to 2 000 m, in areas of thin woodland. At night, like other pheasants, it roosts in the treetops. It is a poor flier, the cocks in particular having difficulties with their long tails. In the courtship season the male makes a chattering call or short chirping sounds to attract the females. Older males have 2—3 hens, but young birds usually have just one. The female lays 7—15 eggs, varying from greenish-brown to cream in colour, and incubates them for 24—25 days. This species feeds and rears its young in the same manner as the Red Jungle Fowl.

50 cm. The female is about 50 cm long, and olive brown in colour, with white lengthwise lines along her back and sides. Her tail is striped with white and chestnut. Young cocks can be recognized by their black and white striped tail feathers. The Mikado Pheasant lives in oak woods or juniper thickets and in bamboo jungles. It forages in the morning and in the evening, consuming a variety of seeds and berries, leaves and grasses in open countryside. Insects and insect larvae are also consumed, especially by young pheasants. The female lays 5—10 eggs in a nest situated under a bush. The eggs are usually yellowish-white in colour, and are incubated by the female for 26—28 days. The breeding season lasts from late February to May.

GOLDEN PHEASANT
Chrysolophus pictus

The Golden Pheasant is one of the most beautiful species of pheasants. The cock reaches a length of 110 cm including the tail, which is up to 80 cm long. The female measures only about 65 cm. She has predominantly greyish-brown coloration with tiny blackish stripes and an orange-brown head. The Golden Pheasant inhabits mountainous areas of central China, occurring in forests up to a height of 2 200 m. It is solitary outside the breeding season, or forms small flocks in winter. In the courtship season, one cock can have as many as 8 hens. The hen makes a shallow scrape under a dense bush, and lines it with a few grass stalks or leaves. Here she lays 5—12 cream to yellow-coloured eggs and incubates them for 22 days. She watches over her brood by herself, for the cock takes no further interest in his family. This pheasant feeds on insects and insect larvae, worms, small young

Golden Pheasant

vertebrates, green shoots, berries and seeds, found on
the ground.

Malaysian Peacock Pheasant

MALAYSIAN PEACOCK PHEASANT
Polyplectron bicalcaratum

The Malaysian Peacock Pheasant is common in forest
areas ranging from the eastern Himalayas to Vietnam.
The cock measures 65—75 cm, while the hen is about
50 cm long. She resembles the male in colour, but is
darker and her tail pattern is less striking. This species lives in dense jungles, usually up to a height of
about 1 200 m. For most of the year it is solitary, but
it forms pairs in the breeding season. The hen makes
a shallow scrape under a bush, lining it with dry
leaves. Here she lays 2—3 pale cream eggs with whitish dots. She sits on the clutch for 21 days, and later
protects the brood by herself. The Malaysian Peacock
Pheasant feeds mainly on small seeds, but it also eats
berries and small invertebrates, particularly insects.

PALAWAN PEACOCK PHEASANT
Polyplectron emphanum

The Palawan Peacock Pheasant reaches a length of
50 cm. The female measures about 10 cm less, and
has a dark brown crest, fawn cheeks and throat, and
brown upper parts with fine, lighter coloured spots.
This species inhabits the island of Palawan, where it
lives in dense forests. Although it abounds in some localities, it hides away in dense cover, and is rarely
seen. The hen makes a shallow depression in thick
grass under a bush, and lines it with leaves. Here she
lays 3—5 ochre-coloured eggs with dark spots, and
incubates them for 21—22 days. This pheasant feeds
on the ground on seeds, berries, insects and insect larvae, and other small invertebrates.

Palawan Peacock Pheasant

GREAT ARGUS PHEASANT
Argusianus argus

The Great Argus Pheasant is a most beautiful bird.
The cock reaches a length of 200 cm including its
long tail feathers, which measure over 120 cm. It also

Great Argus Pheasant

Common Peafowl

has remarkably elongated secondary flight feathers 80—100 cm long. These wing quills are adorned with magnificent, black-rimmed, brilliantly-coloured eyespots. The hen is more drab in colour and only about 75 cm long. The Great Argus Pheasant is found in rainforests from Thailand across Malaysia to Sumatra and Borneo. It lives both on low-lying land, and in mountain areas up to a height of 1 300 m. It is normally found on dry, rocky sites, though it may occasionally be encountered in coastal bamboo thickets. For most of the year it is solitary, but in the breeding season it forms pairs which sometimes stay together after the brood is fledged. Each male prepares his own courtship ground in the middle of an area of dense jungle. This covers 6—8 square metres, and the cock keeps it clear by constantly removing all the vegetation. The courting pheasant stands on the ground or on a low branch which has been stripped of leaves, and lifts his wings obliquely upwards to display the

spots on his secondary flight feathers. He also raises his tail feathers vertically and stretches his head forward, sometimes making loud hooting sounds at the same time. The hen usually lays 2 whitish eggs with buff-coloured spots in a shallow depression under a bush. She incubates them alone for 24—25 days and looks after her chicks. This species feeds on insects and other invertebrates, and on fruit, berries, seeds and shoots. Interestingly, the spectacular wing feathers of this pheasant were already known in Europe in the eighteenth century, while the first live specimen was brought to London in 1872.

COMMON PEAFOWL
Pavo cristatus

The Common Peafowl is native to India and Sri Lanka. The peacock reaches a length of about 220 cm including the elongated feathers above the tail, which can measure over 150 cm. The peahen is smaller, being only 85 cm long, and is brownish in colour, with a green neck and breast. The long, colourful tail feathers are usually shed in July, new ones growing before the winter. Some males do not grow these feathers until they are 3 years old. The Common Peafowl inhabits forests from the lowlands up to heights of 2 000 m. Early in the morning and in late afternoon, the piercing call of the peacock can be heard far away. Surprisingly, these birds can fly despite their long trains, but only for short distances. Peafowl are said to live in the company of tigers, but this is not the case. In India tigers are found in the same localities as peafowl, but whenever peafowl encounter a tiger, they warn the whole neighbourhood. The peacock is polygamous, living with a small group of 4—5 peahens, but he takes no part in nest-building. Each peahen makes a shallow depression in the ground and lines it with grass and leaves. Here she lays 4—8 eggs, yellowish-white in colour, and with fine, irregular dark dots. She sits on the clutch for 28 days and subsequently cares for her offspring. The Common Peafowl feeds on insects and other invertebrates, and on small vertebrates, such as geckos and small snakes. It also eats green shoots, various fruits, grass and seeds.
The Common Peafowl is often reared in captivity, especially in wildlife parks. In India it has been raised since 2 000 B.C. It has been successfully naturalized in all its habitats, but it has never formed a domestic breed and does not differ from the wild form. In an-

cient Rome the tongues and brains of peafowl were cooked with Indian spices and served to Roman emperors as choice delicacies.

PAINTED PARTRIDGE
Francolinus pictus

The Painted Partridge is distributed in central and southern India and Sri Lanka. It reaches a length of 33 cm, and the female is less colourful than the male. The Painted Partridge normally lives in families or singly, forming pairs only in the breeding season. The nest is built under a bush or in a thick clump of bamboo by the female, and she lines the depression with grass. The clutch is composed of 3—7 cream-coloured eggs, sometimes with olive spots, and they are incubated by the female for 22 days. The Painted Partridge feeds on small seeds, berries and green shoots, and on insects and insect larvae. This partridge is a common game bird in its homeland.

BUSTARD QUAIL
Turnix suscitator

The Bustard Quail is a small bird, only about 15 cm long, which closely resembles the quails. The female is slightly larger than the male, and more colourful. This bird is distributed throughout south and southeast Asia, extending to Japan, the Greater Sunda Islands, Sulawesi and the Philippines. It is usually found along the edges of forests, but it can also be seen in parks and in bush-covered localities. In the mountains, it can be encountered up to a height of 2 500 m. It is mostly resident, only leaving moist sites for drier situations in the monsoon season. It prefers thick vegetation, being adept at penetrating it to find shelter in the thickets. When flushed, it hides in the undergrowth and takes to the air only when it is under immediate threat. The Bustard Quail feeds on grass, seeds and green shoots and on tiny insects, particularly termites and ants.

Although nesting can take place at any time of year, it most frequently occurs between June and September. The nest is built in a depression in a tuft of grass or under a low bush, and often it has no lining. During the courtship period, the female fights her rivals for possession of the male. She lays 4 greyish eggs with blackish-brown spots, but takes no further interest in them. The male sits on the clutch for 13—16 days and watches over the young, guiding them to food and protecting them. The female lays several clutches each year and mates with several different males.

Painted Partridge

Bustard Quail

Rufous Dove

Pompadour Pigeon

RUFOUS DOVE
Streptopelia orientalis

The Rufous Dove is widespread from Sri Lanka and India to central and south-eastern Asia and Japan. It is about 30 cm long and both sexes are identically coloured. It lives in thin woodland, on the edges of primary forests and in bush-covered localities. In the Himalayas, it can be encountered at a height of 4 000 m. Rufous doves from more northerly regions migrate to India, large flocks spending the winter there in the scrub. They forage on the ground, collecting seeds and small berries. In the nesting season, each pair builds a nest of dry sticks on a branch or among bamboo stalks, usually 3—4 m above the ground. The female lays 2 white eggs, which are incubated by both partners for 15 days. The young are fed by both parents and leave the nest after 3 weeks.

POMPADOUR PIGEON
Treron pompadora

The Pompadour Pigeon has an area of distribution covering southern and south-eastern Asia, Sumatra, Java, the Moluccas and the Philippines. It is a small

Green Imperial Pigeon

species, reaching a length of only about 26 cm, and it has broad, red feet. The female is less impressively coloured. This fruit-eating pigeon lives in forests, usually in hilly situations at a height of 1 200 m. It feeds mainly on small fruits, especially wild figs, and it often flies long distances to get them. It also eats berries which it pecks from the trees, and it rarely visits the ground. For most of the year it wanders through the forests in small coveys, often raiding fruit plantations.

In the nesting season these pigeons form pairs and build flat nests of small twigs among the greenery of trees or tall bushes. The female lays 2 white eggs, which are incubated by both partners for 12—14 days.

GREEN IMPERIAL PIGEON
Ducula aenea

The Green Imperial Pigeon is distributed in southern and south-eastern Asia, from where it extends to the Philippines and across the Greater Sunda Islands to New Guinea. It inhabits forests at a height of about 300 m, and sometimes visits primary forests at a height of 1 000 m in the mountains. This pigeon can also be encountered in the tall mangroves along the coast. The Green Imperial Pigeon is a sturdy species reaching a length of 50 cm, and the sexes are alike in colour. It feeds on various fruits, particularly figs, and on the reddish berries of nutmeg trees. The pigeon pecks the nutmegs directly from the trees and swallows them whole, although they sometimes measure as much as 4 cm across. The forays for food take place mainly in the early morning and in the evening. Except in the nesting season, the Green Imperial Pigeon travels in groups of 5—6, or sometimes lives singly. It flies down to the ground only to drink from shallow pools. In the nuptial season, it produces laughing sounds similar to the croaking of frogs. The nest is built by both partners 10 m above the ground in a deciduous tree, often a fruit tree in a park or garden. It consists of just a flat platform of dry twigs. The female usually lays a single white egg and both parents share in incubation for 16 days.

EMERALD DOVE
Chalcophaps indica

The Emerald Dove inhabits forests of south and

south-east Asia, the Philippines, the Greater Sunda Islands and Australia. On the southern slopes of the Himalayas, it can be encountered up to a height of 2 700 m. The Emerald Dove is a beautiful bird, about 26 cm long. It flies speedily and expertly among trees, executing sharp turns with admirable skill. It often visits open country along the edges of jungles, or tea plantations, where it forages on the ground. It feeds on seeds, grains and various fruits, especially the berries of the shrub *Lantana aculeata*. It also catches small insects. At midday these doves rest in the tops of the trees. In the courtship season, the males make calls similar to those of cuckoos. The pair builds a flat nest of sticks, 2—3 m above the ground in a dense bamboo thicket within a forest. Breeding takes place throughout the year, according to locality. The female lays 2 white eggs, which are incubated by both parents for 12 days, and both parents feed the young on food regurgitated from the crop.

LUZON BLEEDING-HEART PIGEON
Gallicolumba luzonica

The Luzon Bleeding-heart Pigeon is about 26 cm long. It has a conspicuous, bright red spot on its breast, so that the bird looks as if it just suffered serious injury. The red spot is smaller in the females. This species inhabits the forests of the Philippine islands of Luzon and Polillo, and is abundant in some areas. It lives in pairs, or forms family groups when the broods are fledged. It forages mainly on the ground under trees and bushes, collecting seeds, berries, or occasional small insects, spiders and other invertebrates. The nest is made of twigs and dried grass stalks, and is situated near the ground in a bush or tree. The female lays 2 white eggs, which are incubated by both partners. The young hatch out after 16 days and are fed by both their parents. They leave the nest when they are 14—15 days old.

NICOBAR DOVE
Caloenas nicobarica

The Nicobar Dove is undoubtedly one of the most beautiful species of dove. It is about 40 cm long and has remarkably elongated neck feathers, which form a mane. The female is similar in appearance to the male, but she has a greyer neck and breast. There are

Emerald Dove

Luzon Bleeding-heart Pigeon

Nicobar Dove

Indian Lorikeet or Vernal Hanging Parrot

settle near each other, and form a small colony. There can sometimes be ten nests or more in a single tree. Pairs build their nests from twigs, 3—8 m above the ground in evergreen trees. In the courtship season, the males often make crowing sounds. The female usually lays a single egg, and both parents incubate it for 16—18 days, and rear the chick together.

INDIAN LORIKEET or VERNAL HANGING PARROT
Loriculus vernalis

The Indian Lorikeet, or Vernal Hanging Parrot, is found in an area stretching from India to Thailand and the Andaman Islands. It attains a length of 13 cm. The female is predominantly yellowish-green in colour and lacks the blue throat patch. This small lorikeet lives in woodland up to a height of 2 000 m. Although it is abundant, it is rarely seen because of its secret way of life. However, it does occasionally visit plantations. It lives in pairs which form small flocks when the nesting season is over. Several flocks often gather in the same flowering tree, where the birds feed on nectar, and later on soft fruit or small seeds. These birds are good climbers and often hang on the ends of branches with their heads upside down to reach a blossom. Like its relatives, this lorikeet hangs upside down to rest. The nesting season takes place between January and April. The nest is made in a hollow tree or in a hole in a dead branch, not too far from the ground, and the nesting cup is lined with green leaves. The female lays 2—4 white eggs and incubates them for 19 days, occasionally being relieved for brief intervals by her partner. The young are fed by both parents.

two subspecies, distributed from the Nicobar Islands eastwards across Malaysia, Thailand and Cambodia to the Greater Sunda Islands, the Philippines and New Guinea. The Nicobar Dove is abundant mainly on wooded islands where it lives along the edges of forests. During the day it usually perches in dense foliage in the tops of trees, coming out at dusk, which is its most active period of the day. It runs speedily on the ground, looking for fallen seeds, berries and green shoots, and it pecks up small invertebrates as well. Outside the breeding season, small flocks of 3—30 live together, though many doves remain solitary. They sometimes fly from one island to another in search of ripe fruit, for they are fast, though rather clumsy fliers. In the nesting season, several pairs often

BLUE-CROWNED HANGING PARROT
Loriculus galgulus

Southern Thailand, Malaysia, Sumatra, Borneo and the adjacent islands are the home of the Blue-crowned Hanging Parrot. This small bird is only 12 cm long. The female resembles the male, but does not have a spot on her throat and back. This hanging parrot lives in forests, usually in low-lying land, though it is sometimes found in mountain areas up to a height of 1 250 m. It feeds on nectar, blossoms, fruit and seeds, and on small insects and insect larvae, hanging up-

Blue-crowned Hanging Parrot

side down to feed. The nesting season lasts from January to July. The nest is built in a hole in a dead part of a tree, about 12 m above the ground. The female lays 3 eggs at the bottom of the hole, which is lined with pieces of leaf. While she is sitting on the eggs, the male brings food for her. The young are fed by both parents for 5 weeks on the nest, and for a further 10 days after leaving it.

ROSE-RINGED or INDIAN RING-NECKED PARAKEET
Psittacula krameri

The Rose-ringed or Indian Ring-necked Parakeet occurs in two subspecies in Sri Lanka, India and south-eastern China. This parakeet has also been introduced in Arabia, Iraq and Afghanistan. It reaches a length of about 40 cm. The female lacks the black band on the chin and the red neck band. Outside the breeding season, the Rose-ringed Parakeet lives in small flocks of 10—20. When foraging for food in the fields, the flocks assemble often forming huge congregations of several hundred birds. In the heat of the day, when the parakeets rest in the shade, up to 300 birds may roost in a single tree. Their food consists in the main of seeds, grain, fruits and nectar, though they occasionally eat insects and larvae. In the nesting period, the parakeets form pairs and build nests in hollow trees. The female lays 2—6 eggs, which she incubates alone for 22—24 days, though the male sometimes relieves her for brief intervals. The young birds leave the nest when they are 6—7 weeks old, but they continue to be fed for a further 2 weeks.

LARGE INDIAN PARAKEET
Psittacula eupatria

The Large Indian Parakeet is distributed in Sri Lanka, India and south-eastern Asia. It is probably also found in Pakistan. It reaches a length of 58 cm. The female lacks the black band across the cheeks and the pink neck band, and she is slightly smaller than the male. This parakeet lives in small coveys on the borders of forests, or in large parks and gardens, where it flies among the trees and climbs along the branches in search of seeds, berries, fruit, shoots and nectar. For most of the year, the flocks roam throughout the countryside. Pairs build their nests in tree hollows, espec-

Rose-ringed or Indian Ring-necked Parakeet

Large Indian Parakeet

Blossom-headed or Plum-headed Parakeet

ially in palm trees. The female lays 2—4 eggs and incubates them for 3 weeks, while the male brings her food. The young are fledged after 7 weeks and join their parents on their forays.

*Upper Yangtse
or Chinese Parakeet*

BLOSSOM-HEADED or PLUM-HEADED PARAKEET
Psittacula cyanocephala

The Blossom-headed or Plum-headed Parakeet inhabits Sri Lanka, India, Pakistan, Nepal and Bhutan. It is about 33 cm long. The female has a yellow-rimmed, blue-grey head. This parakeet is a resident of forests at heights from 1 000 to 1 500 m. It lives in pairs, or flies in family groups after the fledging of the young. It feeds on a variety of fruits, seeds, shoots and occasional insects. The nest is built in a tree cavity, where the female lays 4—6 eggs. They are incubated for 22 days, the male also being seen to sit on the clutch. Both partners feed the young for 7 weeks on the nest and for a further two weeks when they are fledged.

UPPER YANGTSE or CHINESE PARAKEET
Psittacula derbiana

The Upper Yangtse or Chinese Parakeet is native to south-eastern Tibet, north-eastern Assam and south-western China. It also inhabits the island of Hainan and northern Vietnam. It is about 50 cm long. The female has a black upper jaw and lacks the blue coloration on the forehead. This species dwells in the dense canopy of the rainforests. It lives in pairs, or in family groups when the broods are fledged. The nest is built in hollows, usually 10—15 m above the ground. The female lays 2—5 eggs and incubates them for 22—25 days, while the male feeds her. The chicks leave the nest at the age of 6—7 weeks. At that time they all have rich red beaks, and the sexes cannot be distinguished, but the beaks begin to turn black within a month, and become completely black after 4 months. At the age of 15—18 months, the beaks of the male birds turn red again, this time permanently, while the females' beaks remain black. The diet of this parakeet is made up partly of seeds, various fruits, shoots and bark, and partly of insects and their larvae.

BEARDED or RED-BREASTED PARAKEET
Psittacula alexandri

The Bearded or Red-breasted Parakeet is widely distributed in the Himalayan region, in southern China, south-eastern Asia, on the Andaman Islands, Java, Bali and Kangean. It has also been introduced in southern Borneo. It is about 33 cm long and occurs in 8

subspecies. This parakeet lives in pairs, or, outside the breeding season, in small flocks of up to 50. It dwells in dense forests, seeking food mainly in the tops of the trees, and feeding on seeds, fruits and shoots. The nest is built in a tree hollow, where the female lays 3—4 eggs and sits on them for 25 days. The young are fully fledged after about 50 days.

ECLECTUS PARROT
Eclectus roratus

The Eclectus Parrot has a wide area of distribution. It is native to the Lesser Sunda Islands, Tanimbar, Kai and Aru Islands and it ranges to New Guinea and north-eastern Queensland in Australia. It reaches a length of 35 cm, and the sexes differ in coloration, the male being predominantly green, while the female is red with a bluish-violet abdomen. There are 10 subspecies. For most of the year these parrots stay in small flocks, forming pairs only in the breeding season. The nest is built in a hole in a tree. The female lays 2—5 eggs, which she incubates for 26 days, while the male regularly feeds her. The young are fed by both parents and leave the nest after 10—12 weeks. This parrot feeds on seeds, nuts, shoots, fruit, berries, insects and insect larvae.

BLUE-CROWNED ROCKET-TAILED PARROT
Prioniturus discurus

The Blue-crowned Rocket-tailed Parrot is native to the Philippines. It reaches a length of 27 cm. The female resembles the male in coloration but has less elongated tail feathers. In the nesting season, pairs

Blue-crowned Rocket-tailed Parrot

Chattering Lory

Red Lory

build their nests near each other, to form colonies. The nests are situated in the trunks of dead trees. Nothing is known of the nesting habits of this species. It feeds on fruit, berries, shoots and seeds.

CHATTERING LORY
Lorius garrulus

The Chattering Lory is indigenous to the Moluccas, where it is a common denizen of the forest canopy. It is about 30 cm long. It usually lives in pairs which form family groups after the fledging of the young. The female lays 2 eggs, in a hollow in a dead tree, and incubates them for 25 days. The young are fed by both parents and leave the nest after 75 days. The diet of this beautiful lory consists of nectar and soft, sweet fruit.

RED LORY
Eos bornea

The Red Lory is a resident of the southern Moluccas and the island of Kai. It reaches a length of 31 cm and occurs in 4 subspecies. Both sexes are identically coloured. This lory frequents large wooded areas and also visits the edges of parks. It feeds on flowers and nectar, and on insects and insect larvae. The female lays 2 eggs in a hole in a tree and sits on them for 25 days, while the male feeds her regularly. The young are fed by both parents and leave the nest when they are 7 weeks old.

LESSER SULPHUR-CRESTED COCKATOO
Cacatua sulphurea

The Lesser Sulphur-crested Cockatoo is found in the Lesser Sunda Islands, Sulawesi and Butung. It is 33 cm long, and the sexes are alike in colour. It is extremely abundant and is often seen in city parks and

Lesser Sulphur-crested Cockatoo

Salmon-crested or Pink-crested Cockatoo

Masked Owl

gardens. Outside the nesting season, it lives in flocks made up of several scores of birds. The nest is built in a hole in a tree, and here the female lays 2—4 eggs, which are incubated by both partners for 25 days. The young are also fed by both parents and leave their nest after 10 weeks.

SALMON-CRESTED or PINK-CRESTED COCKATOO
Cacatua moluccensis

The Salmon-crested or Pink-crested Cockatoo is confined to the southern Moluccan islands of Seram, Saparua and Haruku. It has also been introduced on Ambon. It is a large species, reaching a length of 52 cm. It frequents forests but often visits parks with large trees, in which the bird seeks food. It feeds on fruit, berries, nuts, seeds, leaves and fine bark. The nest is built in a sizeable hollow in a large tree, usually 10 m or more above the ground. The female lays 2 eggs and sits on them alternately with the male for 30 days. The young are fed by both parents, and they leave the nest when they are 6—9 weeks old.

MASKED OWL
Tyto novaehollandiae

The Masked Owl, which attains a length of 35 cm, is found in Sulawesi and it is also common in New Gui-

nea and Australia. In Sulawesi it frequents forests and bush-covered sites, sometimes near human dwellings. It builds its nest in tree hollows or in holes in rocky terrain in the forests, and there are often two broods in a year. The female lays 3—4 white eggs in the bottom of the unlined nest, and sits on the clutch for about a month, while the male brings food to her. The young are cared for by both parents, and leave the nest after about 50 days. This owl feeds on small mammals, birds and small lizards.

BAY OWL
Phodilus badius

The Bay Owl is widely distributed throughout southern and south-eastern Asia and extends to the Greater Sunda Islands. This owl, which is about 30 cm long, inhabits dense primary forests, from low-lying land to mountainous areas at a height of 2 300 m. It is a common species but it is rarely seen on account of its nocturnal existence in the heart of the jungle, where it hunts small mammals, birds, lizards and frogs, beetles and other larger insects after nightfall. In the nesting season, it makes whistling sounds composed of 3 musical notes. The unlined nest is built in a hole in a tree, and here the female lays 2—5 eggs. The nest, which is usually situated 2—6 m above the ground, is often used for several seasons.

MALAY EAGLE OWL
Bubo sumatranus

The Malay Eagle Owl is about 45 cm long. It is found in Malaysia, southern Thailand and the Greater Sunda Islands, where it lives in evergreen primary forests up to a height of 700 m. It dwells on rocky terrain, where it hunts small mammals, lizards, birds, snakes, frogs, or even large insects. It hunts after sunset, hiding by day in holes in trees or in rocky crevices. The nest is built in a hollow by the female, and she lines it with a few feathers from her prey. She lays 2—4 white eggs and sits on them for 35 days while the male brings food near to the nest for her. The young grow rapidly, and take their first flight at the age of 80 days. They soon begin to fend for themselves and seek their own territories. After the young have fledged, eagle owls usually live singly.

REDDISH SCOPS OWL
Otus rufescens

The Reddish Scops Owl is about 18 cm long and its area of distribution covers the primary forests of the Greater Sunda Islands, the Philippines, Malaysia and southern Thailand. It can usually be seen in hilly locations from heights of 300—700 m, and occasionally in the mountains at an altitude of 2 100 m. During the day this owl sleeps in holes vacated by woodpeckers, or among dense branches in the treetops. It comes out to hunt after nightfall, catching airborne beetles, nocturnal moths and other insects, and sometimes seizing small birds. It also occasionally takes small reptiles, particularly geckos climbing in the trees. The nest is built in a hole in a tree, usually about 12 m above the ground. The female lays 2—6 white eggs and incubates them for 26 days. Both parents carefully tend their young, feeding them mainly on moths.

JUNGLE OWLET
Glaucidium radiatum

The Jungle Owlet, which is about 20 cm long, is confined to Sri Lanka, India and western Burma. It dwells in dense woodland but often settles in plantations containing tall trees. In the Himalayas it occurs at a height of 2 000 m. It is predominantly nocturnal, but on overcast days it hunts during the daytime. It is most active an hour before nightfall, and it preys until dawn on grasshoppers, crickets and cicadas, caught on the wing. It occasionally also catches small lizards, birds and rodents. During the day this owlet sleeps always in the same hollow, usually one vacated by woodpeckers or barbets. In the breeding season, which in India takes place from March to May, owlets form pairs. The male at that time often makes a slow, cuckoo-like call which carries over great distances. The nest is built in a hole in a tree usually 3—8 m above the ground. The female lays 3—4 whitish, rounded eggs, which she incubates for 28 days, while the male brings food to her. Both parents feed the young mainly on insects.

RED-FACED MALKOHA
Phaenicophaeus pyrrhocephalus

The Red-faced Malkoha, which is a relative of the cuckoos, is native to Sri Lanka and southern Kerala. This bird is about 50 cm long, and it has an extremely long tail and short, rounded wings. It dwells in rainforests up to a height of 1 700 m. After the nesting season it lives in small flocks, but many malkohas are solitary. The birds hop along the branches in the forest canopy, searching for food which consists mainly of various berries. The malkoha is a poor flier and covers only short distances. In the breeding season, it makes whistling noises. The nest is built in a tall shrub in the forest, and is a bowl-shaped construction in which grass stalks, roots and twigs are intertwined. The female lays 2—3 white eggs. She probably incubates them alone, but little is known about this bird's breeding behaviour. The Red-faced Malkoha rears two broods, from March to May, and from August to September.

Red-faced Malkoha

Red-winged Crested Cuckoo

RED-WINGED CRESTED CUCKOO
Clamator coromandus

The Red-winged Crested Cuckoo is distributed from Sri Lanka and India to southern China and across Malaysia to the Greater Sunda Islands, the Philippines and Sulawesi. It reaches a length of 45 cm and both sexes have the same coloration. This cuckoo inhabits deciduous forests up to a height of 1 600 m, but it also visits city parks and gardens. It lives in small groups of 3—4 birds or singly. It is not timid but prefers to establish itself in the dense canopy. It seeks its food in tall bushes, where it searches for caterpillars, and it also catches butterflies, beetles and other insects. In the nesting season, the males make two-syllable whistling sounds. The Red-winged Crested Cuckoo is a brood-parasite, the female laying her eggs in the nests of other, smaller birds, such as thrushes of the genera *Garrulax* or *Turdoides,* which have pale

Short-winged or Indian Cuckoo

Old World Palm Swift

Crested Tree Swift

blue eggs similar to those of this cuckoo. The thrushes do not usually notice the extra egg in their nest, and proceed to feed and rear the cuckoo's nestling.

SHORT-WINGED or INDIAN CUCKOO
Cuculus micropterus

The Short-winged or Indian Cuckoo has a vast range of distribution in southern, south-eastern and eastern Asia, extending to the Amur basin. It is about 33 cm long and the female is predominantly brown with a light grey throat. The cuckoo leaves the more northerly regions in autumn and migrates to the Greater Sunda Islands and the Philippines where it spends the winter. In its homeland, it lives in dense forests, both low-lying and among the mountains. In Nepal, for example, it can be seen at a height of 3 700 m. It lives singly in the trees, feeding largely on caterpillars, but also catching beetles and other insects. It sometimes alights on the ground, and preys on crickets hidden beneath fallen leaves. The Short-winged Cuckoo is a brood-parasite, the female laying her eggs in the nests of other species of birds. The eggs are either whitish with reddish-brown spots, or they are pale greyish-blue. Each female always produces eggs of the same colour, and she always lays them in the nests of birds which have eggs of a similar colour. Host birds include drongos, minivets, the Black-naped Oriole, and in the Amur region, mostly shrikes. Incubation lasts 12 days, which is 2—3 days less than is required for the host's eggs. As a result the newly-hatched cuckoo usually manages to push out the eggs containing the chicks of its host, and it can then consume all the food brought by its foster-parents.

CRESTED TREE SWIFT
Hemiprocne longipennis

The Crested Tree Swift is an arboreal species found in the Greater Sunda Islands, Sulawesi, Malaysia and Thailand. It reaches a length of 22 cm and has a characteristic crest, erected when the swift is perching. It frequents the edges of primary forests, or large gardens and parks, and in mountain areas, it occurs at heights up to 1 400 m. It prefers forests rich in trees of the genera *Anogeisus* and *Boswellia*. This swift flies in small groups above the trees, preying on various small, flying insects. Unlike its relatives, this species is not an efficient flier, but it moves gracefully, often gliding with stretched wings to the accompaniment of loud cries. It is frequently seen perching on the tips of the topmost branches. It drinks on the wing, cruising above a river or lake and scooping up water with its bill.

The nesting season lasts from March to May, and from July to August. The nest is built on a thin, horizontal, leafless branch. It is a shallow, bowl-shaped structure, made from fine bark, lichen and small feathers, glued together with the swift's saliva. The nest is situated 4—18 m above the ground, and looks like a shoot projecting from a branch, for the dimensions of this extraordinary structure are only 5 × 3 cm, its depth being 10—12 mm. The female lays a single whitish egg with a blue sheen, which completely fills the nest. Both partners share in incubation and in the rearing of the chick. When the swift is sitting on the nest, its breast projects well over the front and its tail feathers protrude at the back.

OLD WORLD PALM SWIFT
Cypsiurus parvus

The Old World Palm Swift measures only 13 cm and has a narrow, forked tail. It is native to Sri Lanka, India, the Greater Sunda Islands and the Philippines. However, it is restricted to localities where there are palms with fan-shaped leaves, such as the Fan Palm (*Borassus flabelifer*), and the Betel Palm (*Areca catechu*). In some areas it is found up to a height of 1 000 m, although it prefers lower situations. It is one of the most common birds of the extensive palm groves. It forms flocks, which move in rather irregular flight, catching insects on the wing, especially winged ants, small beetles and flies. The palm swift often makes a loud, shrill call. It spends the night in an upright position clinging on to the lower part of overhanging palm leaves.

The nesting season takes place in any month of the year, according to locality.The birds build a nest of feathers and fine plant fibres from genera such as *Calotropis* or *Bombax,* glueing the material together with their saliva and fastening the nest underneath the palm leaves. The construction measures about 4 cm across and is 1 cm deep. The female lays 2—3 pure white eggs, which are probably incubated by both partners.

ORANGE-BREASTED TROGON
Harpactes oreskios

The Orange-breasted Trogon, which is 30 cm long, is found in south-eastern Asia and the Greater Sunda Islands. The female differs from the male in having a brownish-grey head and yellowish breast. Trogons live in pairs in dense, evergreen forests, most frequently at heights from 800—1 200 m. After the fledging of the young, they stay for some time in family flocks. Trogons often perch on dry branches with an open aspect. On observing a butterfly or beetle, they take off swiftly and dexterously catch their prey on the wing, before returning to their vantage point. Trogons occasionally peck small ripe berries from trees or bushes, and they fly down to the ground to catch grasshoppers.

The nest is usually built in a hole in a tree or in a depression on the top of a broken trunk, about 6 m above the ground. However, some nests consist only of sticks and twigs, loosely assembled on a branch. The female lays 2—4 eggs, which are usually off-white in colour, and incubation is undertaken by both partners.

Broad-billed Roller or Dollar Bird

Malay Forest Kingfisher

BROAD-BILLED ROLLER or DOLLAR BIRD
Eurystomus orientalis

The Broad-billed Roller, or Dollar Bird, is widely distributed in the warm regions of Asia, from Nepal to Japan, and across the Sunda Islands and the Philippines to Australia. This bird is about 30 cm long, and has a broad, tapering, red bill. It inhabits evergreen forests, living in pairs, or, outside the nesting season, even singly. It often perches for hours on dead branches high in the treetops before taking off to catch an insect on the wing, and returning to its vantage point. It is mainly active in the morning and in the late afternoon, but it often hunts moths in the dark. It occasionally flies down to the ground to catch grasshoppers or crickets, or a small gecko. It is an excellent flier, and in the courtship season it displays and performs aerial acrobatics above the treetops. In the breeding season, it often makes a quiet croaking sound. Rollers nest in different months according to their geographical distribution. They make their nests in hollows in trees which have been abandoned by woodpeckers or barbets, very often in gigantic trees, such as *Bombax* or *Tetrameles,* growing on the edges of a forest clearing. The same hollow is often used for many seasons. The female lays 3—4 white eggs in the hollow and both partners incubate them for about 20 days. Both parents feed the young chicks, which leave the nest when they are about 26 days old.

MALAY FOREST KINGFISHER
Ceyx rufidorsus

The Malay Forest Kingfisher is indigenous to the Philippines, the Sunda Islands, Malaysia and southern Thailand. This small species, which is only about 13 cm long, dwells beside streams in the forests. It perches on branches overhanging the water and waits for prey to appear. As soon as it sees a small fish, it dives into the water and seizes it. It also hunts for frogs, tiny freshwater crabs and insects. For most of the year it roams through its territory and leads a secluded life. It may sometimes fly to mountainous areas up to a height of 1 500 m.

In the breeding season, kingfishers live in pairs. They excavate a burrow up to 60 cm long, in a river bank in the jungle. The tunnel is terminated by a nesting chamber measuring about 7.5 cm. The birds dig with their beaks, and use their feet to remove the soil. The female lays 2—3 white eggs, which she incubates for 18 days, being occasionally relieved by her partner, who regularly brings her food. The young are fed by both their parents in the burrow for 23 days, and for a few days after fledging.

GREAT PIED HORNBILL
Buceros bicornis

The Great Pied Hornbill ranges from south-western China across Thailand and Sumatra. It is a massive bird, up to 130 cm long, the female being smaller than the male, but being similarly coloured. The hornbill dwells in forests up to a height of 1 700 m, but on the Himalayan slopes it can be encountered at an altitude above 2 000 m. It lives in pairs or in family groups of 3—5, which sometimes merge to form flocks of over 30, usually at the time when the birds leave the forests for the nearby fruit plantations. As many as 200 hornbills, gorged with food, can be seen resting on one huge tree after a raid on a banana plantation, so it is no wonder that the growers of tropical fruit regard these birds as pests. In the forests, hornbills feed on figs, nutmegs and various berries. They sometimes eat fallen fruit, or prey on lizards, snakes and small rodents. They also take nestlings. Their powerful bills help them to capture quite large prey.

In the breeding season, the rattling voice of these hornbills is often heard from the heart of the forest. Pairs make their nests in holes in large trees, such as *Cullenia excelsa* or *Calophyllum tomentosum,* the nests being situated 18—25 m above the ground. When she is ready to lay, the female enters the hole and walls up the entrance, leaving only a narrow slit wide enough for her beak to go through. The male helps her from outside. The building material consists of the female's droppings, leaves, mud and sticky pieces of fig brought by the male. The mixture soon becomes as hard as concrete. The male regularly feeds his mate while she lays 2 white eggs and incubates them for 31 days.

At first the newly hatched chicks remain in the nest with their mother, safe from predatory carnivores, raptors and snakes, while the male brings food for them all. When the chicks are about 14 days old, the adults break open the enclosing wall and the female comes out. The hole is then closed again, and both birds feed their young until they are fully fledged and leave the nest. The same hollow is often used for many successive seasons.

WREATHED HORNBILL
Aceros undulatus

The Wreathed Hornbill is resident in rainforests throughout south-eastern Asia, eastern India and the Greater Sunda Islands. The male approaches a length of 115 cm, while the female is slightly smaller and different in colour, being black all over except for a white tail. The Wreathed Hornbill is always found along the edges of evergreen forests, from the lowlands to mountainous areas at a height of 2 400 m. Though abundant, this is a cautious bird, and is seldom seen. Hornbills usually hunt in small groups, but form large flocks when there is an abundance of food in a particular place. The birds eat ripe figs and va-

Great Pied Hornbill

Wreathed Hornbill

Lesser Yellow-naped Woodpecker

rious berries, and catch small lizards, young birds and large insects. They roost in regular sites in the tops of tall trees or in bamboo jungles, arriving shortly before sunset. In the breeding season hornbills live in pairs, and build their nests in tree hollows very high above the ground. Their breeding behaviour is similar to that of the Great Pied Hornbill, the female being walled in while she incubates her 2 eggs.

LESSER YELLOW-NAPED WOODPECKER
Picus chlorolophus

The Lesser Yellow-naped Woodpecker is widely distributed throughout southern and south-eastern Asia, extending to Sumatra and Hainan. It reaches a length of about 27 cm and the female is distinguished from the male by having red patches only on her nape. This handsome woodpecker can be found in forests up to a height of 2 000 m, although it is more common in low-lying and hilly localities. It can also be seen in densely overgrown parks and gardens. This species forms pairs in the nesting season but for the rest of the year it roams the woodland, often in company with other species of woodpeckers, barbets and drongos. It feeds on the larvae of wood-eating beetles and

on tree termites. It breaks up anthills and collects the ants on its long sticky tongue. It occasionally also collects berries.

The nest is built in holes in tree trunks or strong boughs, sometimes only 70 cm above the ground, but sometimes at 20 m. The cavity is usually carved out in a dead tree, and is 30—40 cm deep, with an entrance 5 cm across. The female lays 2—5 glossy white eggs and incubation is shared by both partners for 15—17 days. The young hatch out naked and blind and are fed by their parents mainly on ants and ant pupae. Young woodpeckers leave their hollow when they are 3 weeks old, but they return there to sleep for a few nights before becoming independent.

CEYLON CRIMSON-BACKED WOODPECKER
Chrysocolaptes lucidus

The Ceylon Crimson-backed Woodpecker inhabits forests and extensive mangrove swamps ranging from India across Thailand and Malaysia to the Greater Sunda Islands and the Philippines. It reaches a length of about 33 cm. The female is similar to the male in coloration, but she has a black, white-spotted nape. This woodpecker normally lives on hillsides, usually below 700 m, but it can be seen in the mountains, up to a height of 2 100 m. It forms pairs for life and defends its own territory. The birds are noisy and often fly around to the accompaniment of loud cries. They seek food mainly under the bark of dying trees, pecking out beetles and beetle larvae. They have been seen to catch swarming termites, and they also sip nectar from the flowers of coral, and silk-cotton trees. These woodpeckers drill out a nest 2—5 m above the ground in a tree trunk, often choosing a huge mango tree beside a village or on the edge of a forest. A pair often uses the same cavity for many years, although the birds sometimes carve out a new hole in the same tree. The female lays 2—5 white eggs and takes turns with her partner in incubating them for 14—15 days. The young are fed mainly on insect larvae, and they leave the nest 24—26 days after hatching.

WHITE-BELLIED BLACK WOODPECKER
Dryocopus javensis

The White-bellied Black Woodpecker is native to south-eastern Asia, the Greater Sunda Islands and the Philippines. It is about 50 cm long and the female differs from the male in having a black nape. This woodpecker is a denizen of evergreen forests ranging from low-lying land to a height of 1 200 m. It is also common in bamboo jungles and among tall mangroves in the coastal swamps, and it may take up residence in tall shady trees in tea plantations. These woodpeckers live in pairs and their loud voices are often heard in the vast territories which they occupy. They feed on ants of the genus *Camponotus,* and also on termites and the larvae of wood-eating beetles. They often break into the nests of wild bees of the species *Apis florea,* and as a result some woodpeckers get their head feathers stuck together with honey.

In the nesting season pairs chisel out holes up to 60 cm deep in tree trunks or thick branches. The holes are situated 8—16 m above the ground, and here the female lays usually just 2 eggs and sits on them alternately with the male for 13 days. The young are fed on ants and ant pupae, by both parents.

COPPERSMITH or CRIMSON-BREASTED BARBET
Megalaima haemacephala

The Coppersmith or Crimson-breasted Barbet derives its name from its clear, metallic, monotonous call, often repeated 80 times in a minute, which can be heard from the heart of the rainforest, and which sounds like a hammer beating on metal. This barbet is only 17 cm long. It occurs in forested areas ranging from western Pakistan to south-western China, Sri Lanka, and across south-eastern Asia to Sumatra, Java and the Philippines, and it is often found in gardens and parks in large cities. It is common in all these localities, especially in dry areas. In Nepal and Bhutan, it is found up to a height of 2 000 m. During most of the year this bird moves in groups of 10 birds or so, although a flock of a hundred can sometimes gather in a locality where there is an abundance of food. Barbets feed on ripe figs, berries and various kinds of fruit found in plantations. They also catch termites and airborne butterflies.

The nest is built by both partners in a thick dead bough. Here the birds carve out a hole 25—60 cm long, terminated by a nesting chamber about 15 cm wide. The entrance is about 4 cm across, and the nest is situated from 2—10 or more metres above the ground. Barbets usually settle in trees of the genera *Erythrina, Bombax, Moringa* and *Pongamia.* The fe-

White-bellied Black Woodpecker

Ceylon Crimson-backed Woodpecker

male lays 2—4 white eggs and incubation is shared by both partners. After 16 days the chicks hatch out, and are fed by both their parents.

YELLOW-FRONTED BARBET
Megalaima flavifrons

The Yellow-fronted Barbet, which is about 22 cm long, is native to the forests of Sri Lanka. It mainly inhabits humid low-lying woodland, but it may be encountered in the mountains as high as 2 100 m. It feeds on figs and various berries, often raiding plantations and orchards when the fruit is ripe. Older barbets occasionally catch insects or small tree geckos. The breeding season takes place between February and May and again between August and September. The birds peck out a hollow about 60 cm deep in a dead horizontal branch, and use it over a period of several seasons, each time making it deeper. The nest is situated 2—8 m above the ground, or even higher. The female lays 2—3 glossy white eggs, and these are incubated by both partners for 17 days. The young barbets are fed on soft berries and occasional small molluscs by both their parents.

GREEN BROADBILL
Calyptomena viridis

The Green Broadbill is about 20 cm long, and has beautiful, glossy emerald green plumage. It is resident in forests ranging from southern Burma across Thailand and Malaysia to Sumatra and Borneo, living on hillsides up to a height of 1 000 m. Here its coloration enables it to merge with its background of greenery in the treetops, and so to escape its predators. Broadbills rest during the day, and are active only in the morning and evening, when they forage for food. They feed mainly on berries, but they also catch insects flying past their vantage points on high branches.
Broadbills build very intricate spherical nests from long plant fibres, moss, lichen, pieces of thin bark and leaves. The nest measures up to 1 m, and has a side entrance in the upper part. It is suspended on a thin branch, often above a stream or lake in the forest. The female lays 2—5 white or brownish eggs with rust-coloured spots.

GARNET PITTA
Pitta granatina

The Garnet Pitta, a beautiful bird about 20 cm long,
is an inhabitant of Borneo, Sumatra and Malaysia.
Outside the nesting season, pittas usually live singly
in bamboo jungles or in bush-covered localities in the
forests. They move on the ground, hopping nimbly
among the vegetation and turning leaves with their
beaks in search of food. Pittas prey on beetles, ants
and other insects, and on spiders, worms and tiny lizards. Each bird has its own territory and it defends it
fiercely against intruders. In the evening pittas roost
on the branches of tall trees or bushes.

In the courtship period, pittas form pairs and share in
the building of a huge oval nest shaped like a rugby ball.
This is situated up to 9 m above the ground in a dense
bush or tree. The nest is made of bamboo leaves,
grass stalks and tiny roots, cemented together with
wet mud, and there is a side entrance. The female lays
4—5 brown-spotted eggs, and incubation is carried
out by both partners for 14 days. Both parents care
for the brood, feeding the chicks for 2 weeks in the
nest and for a few days after the young have fledged.

PARADISE FLYCATCHER
Terpsiphone paradisi

The Paradise Flycatcher is found in central, southern
and south-eastern Asia, from which it ranges to the
Greater Sunda Islands. The female reaches a length of
20 cm and has a rust-coloured tail and wings. The
male is pure white except for his black head, and he
has 2 long tail feathers, up to 30 cm long. In flight,
these feathers stream back like a decorative train. In
the subspecies confined to Sri Lanka, the male attains
a length of only 35 cm and has the same coloration as
the female. This flycatcher frequents evergreen forests, parks, large gardens, and mangrove swamps. Its
habitats range from lowland areas to mountains at
a height of 2 400 m. Flycatchers living in northern regions, such as Manchuria or China, migrate south in
autumn. These birds live along the edges of roads and
in clearings, where they perch in trees with an open
aspect. From here they set out to catch insects on the
wing, for flycatchers are swift expert fliers. They also
prey on spiders among the branches.

In the courtship period the male flies among the trees
displaying his tail feathers and singing. From time to
time he perches on a branch,
whirs his wings and repeats
his courting call. The nest is
built 1—15 m above the
ground in a bush or tree. It is
shaped like a deep bowl and
is made of fine grass stalks,
roots, plant fibres and small leaves,
connected with cobwebs. Flycatchers collect cocoons containing spiders' eggs along with the cobwebs,
which makes the nest white on the
outside. The female usually lays 4
eggs, pinkish-white in colour, with
reddish-brown spots. Incubation is
shared by both partners for 15—16
days. The adult birds feed the
young for 2 weeks in the nest and
for a further few days after fledging.

Garnet Pitta

Paradise Flycatcher

Blue and White or Japanese Blue Flycatcher

BLUE AND WHITE or JAPANESE BLUE FLYCATCHER
Cyanoptila cyanomelana

The Blue and White or Japanese Blue Flycatcher is found in the forests of eastern Asia. The female is a brownish colour with a pale spot on her throat. In autumn this flycatcher migrates south to Malaysia, the Philippines, and northern Borneo, being found in wooded areas up to a height of 2 000 m. In spring it returns to its nesting grounds in dense forests. Within two days of their arrival, the males begin to sing, perching in the treetops usually in the morning and evening. The females have also been heard to sing. Female flycatchers build their nests on branches, in holes in trees or in crevices in rocks, using moss torn from the rocks, and incubate their 4—5 white eggs for 15 days. The young are fed on insects and insect larvae by both parents.

SCARLET or FLAMED MINIVET
Pericrocotus flammeus

The Scarlet or Flamed Minivet is indigenous to southern and south-eastern Asia, and it also occurs in the Philippines and the Greater Sunda Islands. It is about 22 cm long. The female is a greyish colour, with a yellow abdomen, yellow wing patches, green feathers underneath the tail, and yellow-edged tail feathers. This beautiful species inhabits woodland up to a height of 2 100 m. The Scarlet Minivet prefers localities close to rivers, but it also takes up residence in parks and neglected gardens. It is rarely seen on the ground, but stays always in the tops of tall trees, seeking caterpillars, beetles, cicadas and spiders among the leaves. For most of the year it wanders through the woods in flocks of about 20 birds.

There are usually two breeding seasons each year. At these times courting males fly up into the sky, spiralling down to perch in the treetops, where they make delicate whistling sounds. Minivets build cup-shaped nests from lichen, moss and cobwebs, covering them with fine pieces of bark from the trees in which they are situated, so that they blend with their background. These nests are usually 6—18 m above the ground in tall trees in tea plantations. The female lines the nesting cup with dry leaves and lays 2—3 pale blue eggs with buff and grey spots. She sits on them for 15 days while the male stays on guard nearby. The young are fed by both their parents.

Scarlet or Flamed Minivet

BORNEO BRISTLEHEAD or BALD-HEADED WOOD SHRIKE
Pityriasis gymnocephala

The Borneo Bristlehead, or Bald-headed Wood Shrike, is a remarkable bird, about 25 cm long, and with bare red skin on its head. It is native to the island of Borneo where it lives in dense inaccessible forests. It spends all its time in the treetops, catching beetles, butterflies and various other insects, as well as spiders and small geckos. It has a crowing voice. Outside the breeding season, it has often been ob-

served in flocks, but nothing is so far known about its
nesting habits, and a nest has never been discovered.

VINOUS-THROATED CROWTIT
Paradoxornis webbiana

The Vinous-throated Crowtit is found in eastern and
south-eastern Asia. It is 13 cm long and has a characteristic short, heavy beak used to peck at bamboo
shoots, berries, seeds, and small insects. For most of
the year crowtits live in groups of 20—30. They wander along the edges of forests at heights up to 2 000 m,
among tall, dense bushes, and in bamboo jungles,
where they also nest. The birds build bowl-shaped
nests, made from stalks, bamboo leaves and cobwebs,
and lined with fine roots. They are situated about
60 cm above the ground in clumps of bamboo stalks.
The clutch consists of 2—4 pale blue eggs.

VELVET-FRONTED NUTHATCH
Sitta frontalis

The Velvet-fronted Nuthatch inhabits wooded areas
up to a height of 2 200 m all over south and south-
east Asia, the Greater Sunda Islands and the Philippines. It is about 13 cm long. The female lacks the
black rim around the eyes. This nuthatch can be seen
in coastal mangrove swamps, or in shady trees in tea
plantations. It usually lives in pairs, but it is sometimes solitary in the non-breeding season. Nuthatches
climb up and down the trees with remarkable speed
and dexterity, prising insects, insect larvae and eggs
from crevices under the bark.
Nuthatches make their nests in holes abandoned by
woodpeckers or barbets, the female usually walling up
a large entrance with a mixture of soil and saliva,
which soon hardens. She lays 4—6 whitish eggs with
fine buffish dots, on a lining of tiny pieces of dry leaf,
and sits on the clutch for 2 weeks. The chicks are fed
on insects by both parents, and leave the nest after
3 weeks to learn to climb and seek insects.

YELLOW-BACKED SUNBIRD
Aethopyga siparaja

The Yellow-backed Sunbird is distributed in southern
and south-eastern Asia, occurring also on the Greater

Borneo Bristlehead or Bald-headed Wood Shrike

Vinous-throated Crowtit

Velvet-fronted Nuthatch

Yellow-backed Sunbird

Black-crowned White-eye

Red-backed Flower-pecker

Sunda Islands, Sulawesi and the Philippines. The male attains a length of 15 cm including the elongated tail feathers, while the female measures 10 cm and has predominantly dark olive upper parts and paler underparts. This sunbird inhabits forests up to a height of 1 300 m, but it is also common in parks and gardens. Outside the nesting season it lives singly in the forest canopy or in tall bushes where it moves with acrobatic skill. It searches for flowers, especially red and pink ones, and sucks out the nectar, often whirring its wings in front of a flower like a large moth as it pushes its tongue into the blossom. It often pecks out a hole at the base of especially large, deep flowers, to reach the nectar, and it also catches small insects and spiders.

The breeding season varies with the area of distribution. A pear-shaped nest with a covered entrance is built from roots, moss, grass stalks and cobwebs, the outer walls being masked with lichens. It is situated on an overhanging branch in a tall bush, or in a bamboo thicket, a few metres above the ground. The female lays 2—3 whitish or cream-coloured eggs with reddish-brown spots. The young sunbirds hatch out after 2 weeks and are fed by both their parents.

BLACK-CROWNED WHITE-EYE
Zosterops atricapilla

The Black-crowned White-eye, a small bird about 10 cm long, is found in south-eastern Asia, the Philippines and Borneo. It inhabits forests, parks and gardens in hilly locations at a height above 300 m, where it lives in the treetops. Here it seeks large reddish flowers on trees of the genera *Bombax, Erythrina* and *Loranthus,* from which it obtains nectar. It also raids fruit plantations, pecking holes in ripe oranges and mangoes and sucking out the juice, and it also feeds on small insects. Outside the nesting season, white-eyes live in groups, often in the company of other insectivorous birds. The flocks fly from tree to tree, making delicate chattering noises. White-eyes are fond of bathing and gather regularly at pools and shallow water tanks.

A small, hemispherical nest is built by the female, from fine grass, roots, moss and cobwebs. It is situated in a horizontal forked branch and hidden among the foliage, about 1—6 m above the ground, and the nesting cup is lined with animal hair. The clutch of 2—4 pale blue eggs is incubated by both partners for

10—11 days, and both parents participate in feeding the young.

RED-BACKED FLOWER-PECKER
Dicaeum cruentatum

The tiny Red-backed Flower-pecker is only about 7 cm long. It is distributed from northern India across all of south-eastern Asia to Sumatra and the Moluccas. The female differs from the male in coloration, being predominantly grey with a red rump. The flower-pecker frequents forests, gardens and city parks, and even dwells among the mountains up to a height of 1 400 m. It lives always in the treetops, or in tall flowering shrubs. Outside the breeding season it roams singly or in pairs, or sometimes forms small flocks. It is extremely agile, being always on the move, and hopping from branch to branch as it seeks small insects and spiders. It also sips nectar and the juice of overripe fruit. It often pecks berries of parasitic plants of the genus *Loranthus,* allied to the English mistletoe, swallowing them whole. The seeds pass through the bird's digestive tract and then become attached to branches, on which they quickly sprout. In this way, the parasitic plants are easily propagated.
The female builds a pouch-shaped nest at the tip of a branch. It is made from plant fibres, fine grass and cobwebs, and the walls are covered with a protective layer of bark and moss. As the entrance is usually hidden under a branch, the nest is perfectly concealed. The nest is situated among leaves 2—15 m above the ground, so it is almost invisible to the flower-pecker's enemies. The female lays 2—3 white eggs with buff spots. Both parents carry out incubation for 10—11 days, and both feed and rear the brood.

GOLDEN-FRONTED LEAFBIRD
Chloropsis aurifrons

The Golden-fronted Leafbird is about 20 cm long, and has most beautiful plumage. Its range of distribution stretches from India and Sri Lanka across south-eastern Asia to Sumatra. It is resident in forests, being found in the Himalayas up to a height of 1 800 m, but it often visits gardens when the flowers are in bloom, and fruit plantations, where it sips the juice of overripe oranges or mangoes. It prefers nectar from flowers of the trees of the genera *Erythrina, Bombax* and

Golden-fronted Leafbird

Loranthus. It also pecks berries and preys on insects and spiders on the branches, often hanging upside down, and performing various acrobatics. The leafbird is abundant in some areas, but its secret way of life in the treetops makes it difficult to observe. Its melodious, whistling call can sometimes be heard in the forests, and it often imitates the song of other birds. It usually lives in pairs. In the breeding season, between May and August, it nests on the tips of overhanging branches. The nest is a bowl-shaped structure made from fine twigs, grass stalks, leaves, moss, and cobwebs. It is situated 9—12 m above the ground and is lined with soft grass. The female lays 2—3 pale yellow eggs with reddish dots. The young hatch after 12 days and are fed by both parents.

FAIRY BLUEBIRD
Irena puella

The Fairy Bluebird, which reaches a length of 27 cm, lives in the forests of south and south-east Asia, the Greater Sunda Islands and Palawan. The female is predominantly dull peacock blue in colour, with black

Fairy Bluebird

Black Drongo or King Crow

White-rumped Shama

primaries and black-tipped tail feathers. Outside the nesting season, the Fairy Bluebird roams in small groups of 6—8, flying among the tops of the highest trees. It hops from branch to branch, pecking berries and soft fruit, or preying on insects, and it sips nectar from flowers, especially of coral trees and silk oaks.

In the breeding season pairs form and seek out a suitable nesting site in the dense, moist forests. They build their nests on branches, often more than 5 m above the ground. The nest is made of dry twigs, about 17 cm long, on which rests the nesting cup, which is made from moss, roots and leaves. The nest is built solely by the female, though the male accompanies her and protects her throughout. She lays a clutch of 2—3 olive grey to greenish-white eggs with irregular brown spots, and incubates them for 15 days. The chicks are fed by both their parents.

BLACK DRONGO or KING CROW
Dicrurus macrocercus

The Black Drongo, or King Crow, is distributed from Iran across India and Sri Lanka to south-eastern Asia and Java. It reaches a length of about 30 cm and the sexes are alike in coloration. It inhabits forest margins, gardens and parks, and it also dwells in bush-covered sites containing scattered trees. It can be found from low-lying situations up to a height of about 2 100 m. It keeps to the treetops, from where its loud, melodious call can often be heard. It feeds on insects, small lizards and nestlings, and in the evening, it even pounces on small bats. It sometimes sips nectar from flowers of silk-cotton trees, such as *Bombax insigna* or *Bombax malabarica*.

Pairs build their nests on deciduous trees, usually on mangoes, mesquites, acacias and oaks. The nest is a bowl-shaped structure situated in the fork of a branch, 2—12 m above the ground. The birds construct it from twigs and plant fibres, adding cobwebs for strength. The female lays 2—3 eggs, whitish or pink with brown spots. The partners take turns at sitting on the clutch for 16 days, and both feed the young. The Black Drongo often rears a young cuckoo in its nest.

WHITE-RUMPED SHAMA
Copsychus malabaricus

The White-rumped Shama is widespread from India

Magpie Robin

and Sri Lanka across the whole of south-eastern Asia to the Greater Sunda Islands. This bird is especially remarkable for the beauty of its song. The male reaches a length of 27 cm, while the female, which is greyish-brown above and pale brown below, measures about 23 cm. The White-rumped Shama lives in thin woodland and abounds in parks and gardens. Although it is usually resident in low-lying land it can also be found up to a height of 1 000 m. In the morning and evening in particular, the male sings a variety of loud, clear melodies, and is often kept as a pet for this reason. Shamas nest in shrubs and bamboo thickets, or sometimes in tree hollows vacated by woodpeckers. The female builds the nest while the male brings stalks, roots, moss and wool to her. She lines the nesting cup with hair, feathers or coconut fibres, and then lays 4—5 dark green eggs, densely spotted with brown, incubating them mostly alone for 11—13 days. Generally the males feed their mates during this time. The newly hatched chicks are usually fed by the female, but their father often joins her, and sometimes he takes over their feeding altogether. The young start hopping out of the nest when they are 12—13 days old and take their first flight a few days later. This bird feeds on a variety of insects, spiders, other invertebrates, and berries. In the breeding season, the male has the strange habit of striking his wings together over his body as he flies.

MAGPIE ROBIN
Copsychus saularis

The Magpie Robin ranges from western Pakistan across southern China to the Philippines, and across south-eastern Asia and the Greater Sunda Islands to the island of Bali. This bird is about 22 cm long, and the female is more subdued in colour than the male. The Magpie Robin abounds in gardens, parks, and along the edges of forests, and in the Himalayas it can be found up to a height of 2 000 m. In the nesting season, the male vigorously defends his territory against intruders, and after nightfall he often calls from the top of a tall tree. His melodious whistling voice can also be heard before dawn. The Magpie Robin sometimes builds its nest in bushes or among the branches of trees, but it also nests in holes in trees or in crevices in walls. The female uses dry stalks, various fibres, roots, twigs, leaves and animal hair to make the bowl-shaped nest, which is situated 1—18 m above

the ground. She lays 3—6 pale green eggs with blackish-brown and purplish-grey spots, and incubates them for 13 days, while the male feeds her. At first the newly hatched young are fed by their mother, but later the male helps her. The chicks leave the nest after 14 days, but their parents continue to feed them for a further 2 weeks. The Magpie Robin forages mainly on the ground, preying on insects, earthworms, molluscs and occasional lizards. It also takes nectar from flowers of coral and silk-cotton trees.

COMMON TAILORBIRD
Orthotomus sutorius

The Common Tailorbird, a tiny bird measuring only about 11 cm, inhabits southern and south-eastern Asia as far as Java. During the breeding season, the

Common Tailorbird

Baya Weaver

Baya Weaver

central tail feathers of the male are 3.5 cm longer than those of the female. This bird is common in bamboo thickets, woods and gardens, and in plantations of tall trees. In the mountains, it can be seen at a height of 1 400 m, or even up to an altitude of 2 000 m. It lives in pairs in the tops of trees and shrubs and in bamboo jungles, sometimes flying down to the ground, where it hops about with its tail held erect over its back. The Common Tailorbird preys on small insects and insect larvae, and on small spiders, and feeds on nectar from coral and silk-cotton trees. It is extremely agile, and its soft, rapid call is a familiar sound in gardens and villages.

Mating takes place in any season, but it usually occurs in the monsoon period when the leaves are at their largest, for trees with huge leaves, such as mangoes, figs, or ornamental garden trees, provide the tailorbird with material for its highly intricate nest. First the bird pushes the edges of one large leaf or several smaller leaves against each other, using its beak and feet. Then it pierces the leaves with its beak, pulling a long plant fibre, or thread of silk from a cocoon or spider web through the holes, and making stitches like a tailor. In this cup the bird makes a nest of reed wool and mammal hair, often lined with horsehair. The nest is usually within 1—5 m of the ground, though it may be higher. The female lays 3—5 eggs, varying in colour from green or bluish, to pink, all with large buff spots. Both partners incubate the clutch for 12 days, the female spending more time on the nest, while the male feeds her. The young are fed by both their parents. The Common Tailorbird often rears a young cuckoo in its nest. Sometimes nests and whole broods of chicks are destroyed by small carnivores, lizards and birds of prey.

THREE-COLOURED MANNIKIN or
BLACK-HEADED MUNIA
Lonchura malacca

The Three-coloured Mannikin, or Black-headed Munia, is about 11 cm long. It is distributed in several subspecies throughout south and south-east Asia and on the Greater Sunda Islands. It inhabits thin woodland, plantations, bamboo thickets and tall shrubs. In Sri Lanka and southern India, it occurs in mountain areas up to a height of 2 100 m, on grass-covered sites. Flocks of mannikins often visit rice paddies, for they feed mainly on grass seeds, green plants and insects. Outside the nesting season, mannikins roam the countryside in flocks of over a hundred, often in company with weavers. The breeding season coincides with the monsoon period, which in southern India lasts from December to March. The nest is built in a shrub or on bamboo stalks, not far above the ground. It is a large, spherical structure, made of grass and other plant fibres, and it has a side entrance. The female lays 4—6 white eggs, which are incubated by both partners for 11—13 days. The young are fed by both parents, and leave the nest when they are 16—18 days old.

BAYA WEAVER
Ploceus philippinus

The Baya Weaver ranges from western Pakistan to south-western China, and across south-eastern Asia to Sumatra and Java. It reaches a length of 15 cm. In the non-breeding season, both sexes have brownish plumage dark-spotted above, and brown beaks. This weaver lives along the edges of forests, usually near rice paddies. In Sri Lanka, it occurs up to a height of 900 m, and in the Himalayas up to 1 400 m, but in winter it leaves the mountain areas for the lowlands. It often visits grasslands containing tall trees, such as acacias or palms. It is a gregarious bird, living in large flocks and breeding in colonies of several hundred. Over 200 nests are sometimes built in a single large tree, mostly of a massive species, such as acacias, rosewoods and mesquites. Mating usually takes place twice a year, most frequently in the monsoon season.

Red Avadavat

The pouch-shaped nest is 60—120 cm long, and hangs from a branch, 3 m above the ground. The narrow upper part is attached to the branch by means of a short plaited rope. The wide central part contains the nest chamber, and the bottom of the nest consists of a long, tubular tunnel, open at the end, through which the birds fly in and out. The nest is woven from thin grass stems or narrow strips of palm leaf skilfully torn off by the birds on the wing. The chamber is lined with small pellets of fine mud, the weight sometimes tearing the nest open. The nest is built entirely by the male, who then constructs 2—4 other nests for subsequent mates. The female lays 3 white eggs and sits on them for 14—15 days. She feeds the young on caterpillars, spiders, rice grains and various other seeds, the male rarely helping to rear the brood. The young leave the nest 15—17 days after hatching but remain in the colony. Juvenile males mate when they are 2 years old, and are fully coloured, while the females can breed at one year of age. The diet of the Baya Weaver consists in the main of seeds, grain, insects and other small invertebrates. The birds often sip nectar from the flowers of coral and silk-cotton trees or of caper bushes.

RED AVADAVAT
Amandava amandava

The Red Avadavat, one of the most popular cage birds, occurs in several subspecies in south and south-east Asia, Java and the Greater Sunda Islands. It has been introduced in many other regions, such as Sumatra and Luzon, where it soon became acclimatized. The Avadavat is a resident of forests, bamboo jungles and scrubland, and it can sometimes be en-

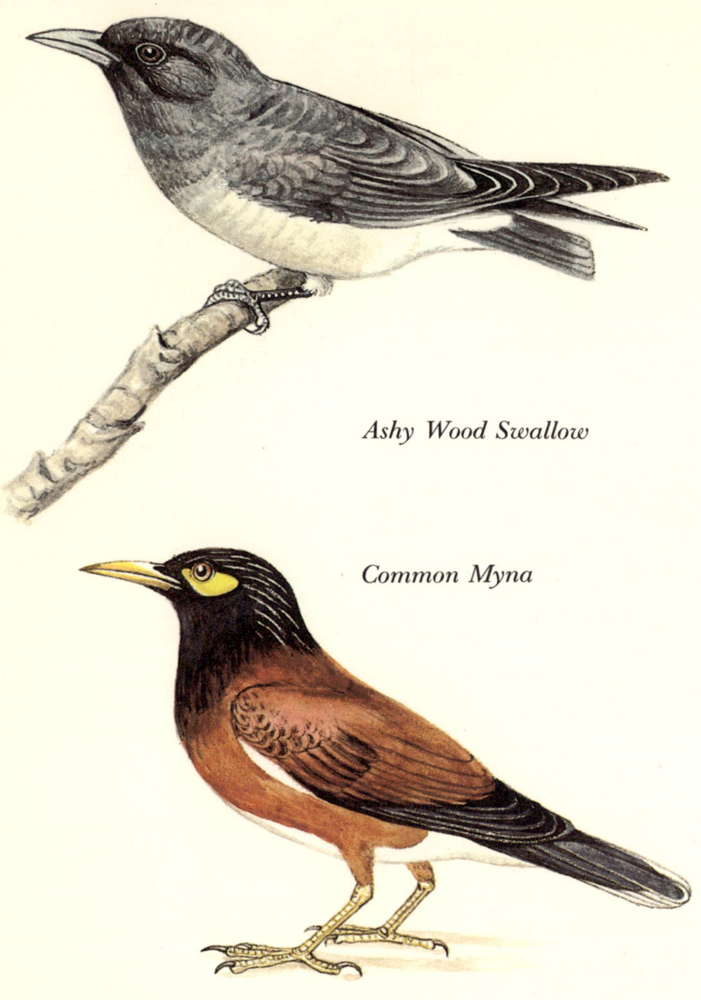

Ashy Wood Swallow

Common Myna

a tunnel. The female usually lays 4—6 white eggs, though sometimes there are up to 10 eggs in the clutch. She mainly incubates them alone, although her mate sometimes relieves her. The young hatch after 12—13 days, and are fed on grass seeds and small insects by both parents. They are fully fledged at the age of 16 days.

ASHY WOOD SWALLOW
Artamus fuscus

The Ashy Wood Swallow inhabits south and south-east Asia. It is about 19 cm long, and resembles true swallows in gracefulness of movement, but it has a straight tail and triangular wings. It lives in thinly wooded forests, on the fringes of jungles, in parks and gardens, and in mountains up to a height of 2 300 m. It usually forms pairs, but often hunts in small coveys, always on the wing. It is an accomplished flier, catching butterflies and dragonflies high in the sky. It also preys on swarming termites, and occasionally feeds on nectar from the flowers of coral trees. It roosts in palm trees and pairs often build their nest at the base of a fan-shaped leaf. The construction is bowl-shaped and made from roots, fine grass and plant fibres, and it is lined with feathers. The female lays 2—3 off-white eggs, with brownish spots. Both partners share incubation for 17 days and both feed the young.

countered in mountains up to a height of 2 100 m. It is especially abundant in tamarind groves. Outside the breeding season it roams the countryside in small flocks of 30 or so, seeking grass seeds, often together with other species of seed-eating birds.

Avadavats nest in trees or shrubs. The nest is constructed mainly by the male, and is a spherical grass structure with a side entrance, generally in the form of

COMMON MYNA
Acridotheres tristis

The Common Myna is distributed from Afghanistan to Sri Lanka, and it occurs in southern Russia. It has also been introduced and become acclimatized in Australia, New Zealand, South Africa and many other regions. It is one of the most plentiful birds in its native habitats. It reaches a length of about 25 cm. It lives along the edges of forests, in parks and gardens, and on mountain slopes up to a height of 1 600 m, although in the Himalayas it can be seen as high as 3 000 m. In winter it leaves these higher localities for low-lying land. It often occurs near human settlements. Outside the nesting season it moves in large flocks of up to several hundred. Every day, before sunset the birds return to their regular sites in the crowns of tall deciduous trees to roost. One after another the flocks arrive at the night quarters, and

Hill Myna

their chattering cries can be heard until the birds fall asleep. Some mynas spend the night under cover of thick bamboo. In the morning when the sun rises , the flocks begin to forage. They feed on seeds, soft fruits, especially figs, jujubes and mulberries, and on nectar from flowers. They also catch locusts, beetles and small rodents, making long forays into the steppes for food.

Mynas form pairs in the breeding season and nest in a hole in a tree, in a cavity in a rock or even in a crevice in the brickwork of a city building. The nest is made from small twigs, grass stalks, roots and feathers, and is used for several seasons. The female lays 4—6 bluish-green eggs, which are incubated by both parents for 17 days. The young are fed by both birds and become fully fledged at the age of 22—24 days.

HILL MYNA
Gracula religiosa

The Hill Myna is widely distributed throughout south and south-east Asia, and is found on the Greater Sunda Islands and on Palawan. It is 30 cm long. This bird inhabits forests up to a height of 1 200 m, or even 2 000 m in the Himalayas. It usually lives in small groups of 5—8, wandering in the forest canopy or among tall bushes covered with ripe berries. Many flocks may gather in forests where figs are ripening. Here they feed, along with barbets, hornbills and fruit-eating pigeons, and as they feed they make a variety of loud noises, some harsh and some melodious. In addition to fruit and berries, they consume a variety of seeds, such as those from the Magnolia (*Michelia champaca*), but they also enjoy nectar from coral trees and silk oaks or from a forest bush *Helicteres isora*. They occasionally catch insects or take small tree lizards.

Pairs of hill mynas often stay together for life. They nest in a hollow vacated by woodpeckers, usually 10—15 m above the ground in a tree growing near the edge of a wood or in a tea plantation. The nest is made of fine twigs, moss, leaves, grass stalks and feathers. The female lays 2—3 pale blue eggs with russet and chocolate-coloured spots, and both partners sit on them for 16 days. The young are fed by both parents.

The Hill Myna is well-known for its ability to imitate voices and sounds, and in captivity it can even learn to speak whole sentences. In the wild it produces vari-

Black-naped Oriole

ous whistling, screeching or croaking sounds, or repeats the calls of other species of birds.

BLACK-NAPED ORIOLE
Oriolus chinensis

The Black-naped Oriole inhabits the forests, parks and mangrove swamps of eastern and south-eastern Asia, the Greater Sunda Islands, Sulawesi and the Philippines. It is about 27 cm long, and the female resembles the male, except that her back is yellowish-olive in colour. In winter, the Black-naped Oriole can be seen in large numbers in India, and the populations from China migrate to Sri Lanka. In the eastern Himalayas, this bird can be found up to a height of 2 000 m. It lives in deciduous forests, where it forages in the tops of the trees for insects, fruit and berries, especially of the tree *Trema orientalis*. It also sips nectar from the flowers of coral and silk-cotton trees.

In the nesting season, the male produces a melodious, whistling call. The birds weave a cradle-shaped nest from thin stalks and bark fibre, in a fork between two twigs on the end of a branch 6—20 m above the ground. The upper rim of the nest is firmly attached to the fork. The female lays 2—3 pink eggs with dark red, violet and grey spots, and incubates them for 14—15 days, being occasionally relieved by the male. The young are fed on insects and spiders by both their parents, and leave the nest when they are two weeks old.

MAROON ORIOLE
Oriolus traillii

The Maroon Oriole reaches a length of 28 cm and is one of the most beautiful species of oriole. The female resembles the male, but she has a brownish-black head, a dark reddish-brown mantle, and a whitish

Maroon Oriole

from the Himalayas to south-eastern Asia and to Borneo and Sumatra. It reaches a length of 40 cm and both sexes are alike in colour. It is found in the tops of tall trees in humid evergreen forests, sometimes up to a height of 2 000 m, but usually keeping below 1 500 m. Outside the breeding season it wanders in pairs, singly, or in small groups, often in company with drongos. It is very abundant, but more heard than seen because it is shy and wary, and conceals itself expertly in the greenery. In contrast, it is an extremely noisy bird, sometimes making melodious whistling sounds, but often imitating the calls of other birds. It is very active, always foraging for insects, preying on small lizards, tree frogs and small birds, or plundering birds' nests. This quality is exploited in the Himalayas where young green magpies are caught, and taught to hunt small birds.

The mating season varies from area to area, but mostly takes place between January and March. These magpies build their nests on top of tall shrubs and trees. They resemble crows' nests, being assembled from twigs, bamboo leaves, moss and roots, and lined with lichens. The female lays 3—5 whitish, brown-spotted eggs. The young hatch after 17 days and receive food from both parents.

throat, breast and abdomen covered with dense black spots. The Maroon Oriole inhabits wooded hills in south-east Asia and in the eastern Himalayas, where it occurs at a height of 2 400 m. It keeps to the treetops, living usually in pairs, or sometimes singly outside the nesting season. It forages among the trees for wild figs, berries, insects and nectar. The nest is built by both birds among the branches, from 4—10 m above the ground, and consists of a deep basket made of plant fibres and cobwebs. The female lays 2—3 pale pink eggs with black and reddish-brown spots, and incubates them for 15 days, occasionally being relieved by the male. The young are fed by both parents in the nest for 2 weeks, and for 10 days after fledging.

GREEN MAGPIE
Cissa chinensis

The Green Magpie is distributed in an area ranging

RED-BILLED BLUE MAGPIE
Urocissa erythrorhyncha

The Red-billed Blue Magpie has a characteristic, 50 cm-long tail, while its total length averages 70 cm. It inhabits humid mixed forests from the Himalayas to south-eastern Asia. It usually frequents mountain areas at heights of 800—1 500 m, but in the Himalayas it may ascend up to an altitude of 2 100 m. Outside the nesting season it wanders in flocks, often with

Green Magpie

other birds of the same family. It generally makes crowing sounds, but it can also imitate the calls of other birds. It seeks its prey, mostly insects and small vertebrates, in the trees, and seldom visits the ground. In the Himalayas the nesting season takes place between April and June, while in Burma it occurs between March and April. The birds build bowl-shaped nests 6—8 m above the ground, using twigs, roots and leaves. The female lays 5—6 eggs, which are off-white in colour, densely covered with brown and buff spots. Both parents share incubation for 17 days and later both feed their young.

KUHL'S GECKO
Ptychozoon kuhli

Kuhl's Gecko, sometimes known as the Flying Gecko, is an unusual lizard found in Malaysia. It has wide flaps of skin on each side of the body, which, when extended, enable the gecko to glide. These gliding organs roll up and the two flaps fold on the gecko's abdomen, like a waistcoat. The gecko has two similar, smaller outgrowths on each side of the head, and the tail is ringed, flattened, and constantly spread. The toes are webbed in order to facilitate gliding movements and they have suckers. The gecko is not capable of active flight, but when it jumps from a branch it stretches out the membranes and comes gliding down. The female lays hemispherical eggs in pairs on the bark of the tree in which she lives, sometimes laying them high in the treetops, beneath orchids growing there. She always lays in the same spot, attaching the new eggs to the remains of the old hatched ones. The shells are soft at first, but they soon harden in the air. The young hatch after 10 weeks, and remain on the same tree throughout their lives, leaving it only when serious danger threatens. The coloration of the gecko, consisting of lighter and darker patches of brown, enables it to remain invisible to its enemies as it clings to the bark. Kuhl's Gecko feeds on small insects, insect larvae and spiders.

COMMON FLYING LIZARD
Draco volans

The Common Flying Lizard, also known as the Flying Dragon, is a widely distributed and abundant species, found in Indonesia and in the south of the Malay Pen-

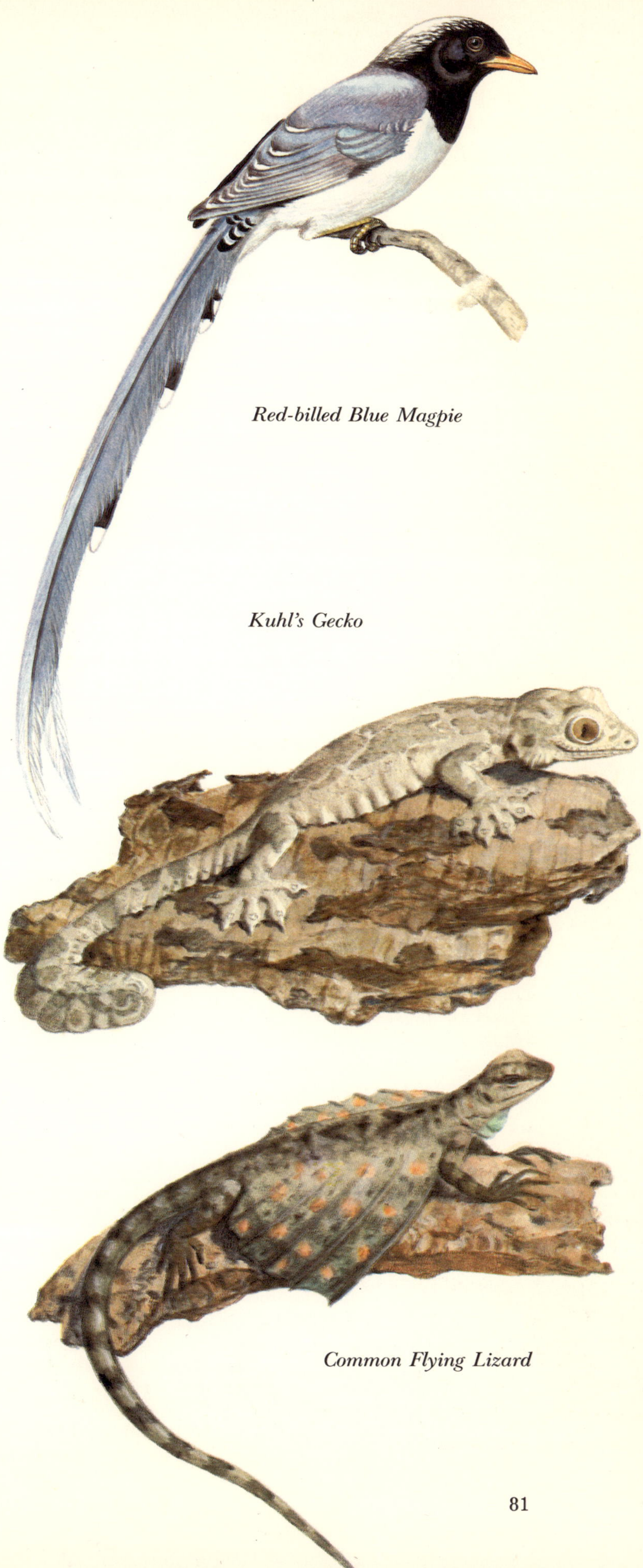

Red-billed Blue Magpie

Kuhl's Gecko

Common Flying Lizard

Variable lizard
Calotes jubatus

insula. This lizard reaches a length of about 20 cm. On each side of the body 5 or 6 long ribs grow out and support a stretchable membrane between them. When at rest the ribs and membranes lie folded along the sides of the body like a closed umbrella, but when this lizard wants to reach another tree it jumps into space, spreads its membrane wide and glides, for all the world like a giant butterfly, with spread 'wings'. The female is brownish, with dark spots, and her flying membrane is yellowish above, and patterned with dark spots. These lizards also have a throat pouch which they can expand and contract. That of the males is an intense orange colour, while the female's pouch is blue or green. The male agitates his throat pouch and spreads his 'wings' when he encounters a rival and he displays them to females in courting rit-

Blind snake Typhlops lineatus

uals. Flying lizards blend easily with their green background, only betraying their presence by spreading their colourful membranes, and suddenly showing the bright orange colour. In this manner males signal to each other that a territory is occupied. Flying dragons prey on insects and insect larvae.

VARIABLE LIZARD
Calotes jubatus

The variable lizard *Calotes jubatus* is a resident of the fringes of forest parks in south and south-east Asia. It is about 40 cm long and is able to change colour very quickly. The males often change colour when they are fighting their rivals, displaying many rich shades which turn pale and dull in the loser of the contest. This lizard abounds in some localities, where it can be seen basking on branches or rocks. It climbs skilfully on rocky walls and on trees and shrubs, preying on insects and larvae, spiders, and even on other small lizards. The female lays 6—12 eggs with membranous coats in a shallow depression, which she digs for herself.

BLIND SNAKE
Typhlops lineatus

The blind snake *Typhlops lineatus* inhabits Indonesia, the Malay Peninsula, Vietnam and southern China. It reaches a length of about 50 cm. It is found in abundance on the verges of primary forests, living in humous soil. It seeks its prey in the evening, visiting termite mounds and anthills, but also feeding on other small insects and on worms. At this time it is often seen as it crawls across a road. The natives are afraid of this blind snake, for they think it is a highly aggressive and dangerously venomous reptile. However, it is harmless and quite unable to bite a man, for it has very small jaws which do not open very wide. The female lays her eggs under pieces of bark or in underground tunnels.

RED-TAILED CYLINDER SNAKE
Cylindrophis rufus

The Red-tailed Cylinder Snake is a strange reptile,
with a cylindrical body and abdominal scales which
are not enlarged as in most snakes. Its skull bones are
also firmly connected. It lives in Burma, Cambodia,
Thailand, Malaysia and Indonesia, and reaches
a length of 80 cm. Its haunts are underground, in the
dens of small mammals, or beneath the fallen trunks
of large trees. It feeds predominantly on the young of
small lizards. When alarmed or molested, it conceals
its head and raises its tail, revealing the crimson-red
or rich orange colour of the underside. The tail then
looks like a head, poised for attack, and this often deters the predator.The female gives birth to 5—10 live
young.

INDIAN SNAIL-EATER
Aplopeltura boa

The Indian Snail-eater is an interesting reptile with
characteristic long teeth. It is found in Indonesia,
from where its area of distribution extends to Malaysia and the Philippines. It reaches a length of 75 cm,
and remains always in the branches of trees and
shrubs, where it forages for snails. Its habitats are
mostly in mountainous areas, usually at heights of
800—1 000 m. This snake is common on the verges
of primary forests, in wet, swampy situations, or near
mountain streams where snails abound. Many snails
climb into the treetops, which are an ideal hunting
ground for this snake, which is strictly specialized in
its diet. It is harmless to man.

GOLDEN TREE SNAKE
Chrysopelea ornata

The Golden Tree Snake lives in Sri Lanka, southern
India, Burma, Malaysia, Indonesia and the Philippines. This slender tree snake reaches a length in
excess of 1.5 m, and is variable in coloration. It is also
known as a flying snake on account of its ability to
glide from branch to branch with considerable speed.
Its abdominal scales are adapted for this purpose, being situated transversely and overlapping each other.
The snake can draw in its abdomen to form a deep
groove which acts as a kind of parachute. The flying

Red-tailed Cylinder Snake

Indian Snail-eater

Golden Tree Snake

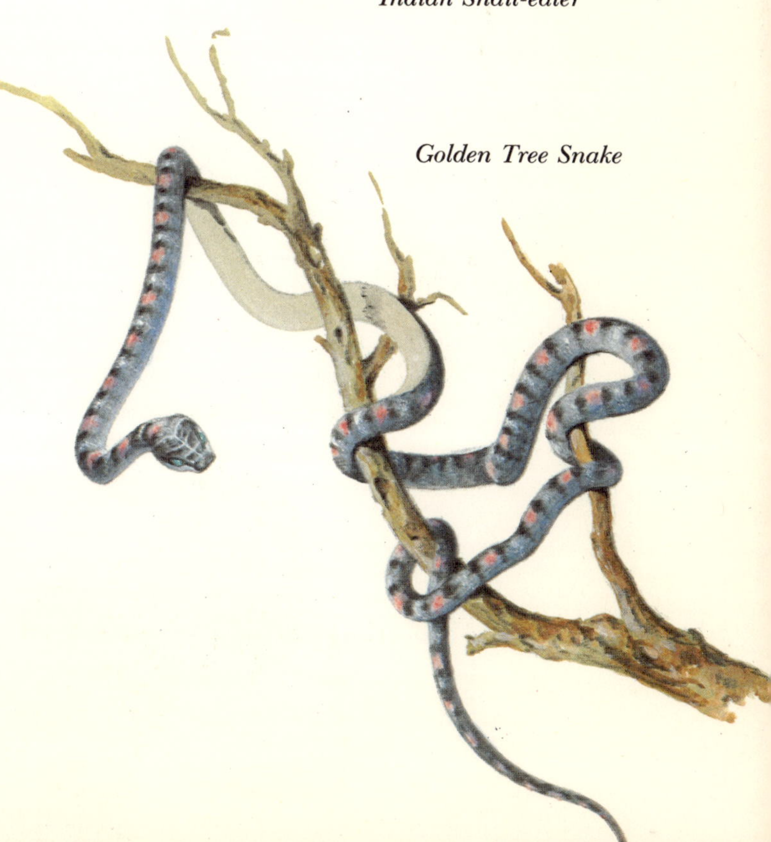

Whip snake
Ahaetulla prasinus

snake slips from a tree and flies like an arrow to a lower branch, not too far away. It is a diurnal reptile, adeptly climbing in the treetops and shrubs as it hunts for various sorts of lizard, and tree frogs. It seizes the prey with its jaws and bites it with the enlarged rear venom fangs. The poison paralyzes almost immediately, so that the snake can swallow the animal easily. The venom is not especially poisonous to man.

WHIP SNAKE
Ahaetulla prasinus

The whip snake *Ahaetulla prasinus* is distributed throughout south-eastern Asia and Indonesia. It is up to 2 m long, and has a body which is extremely thin

Mangrove Snake

and whip-like. It lives in the trees, perfectly blending with the green background because of its protective coloration. It catches small lizards and tree frogs, its long tongue often protruding for a considerable time as it forages in the trees. On seizing its prey, the snake immediately attempts to reach the body of its victim with its venom fangs, which are situated in the rear part of the upper jaw-bone. The venom quickly kills small animals and paralyzes larger ones. The female gives birth to live young, which are active from the beginning and straightaway climb up the trees.

MANGROVE SNAKE
Boiga dendrophila

The Mangrove Snake is a robust reptile, reaching a length of more than 2 m. It inhabits the forests of Malaysia, Thailand, southern Vietnam, the Greater Sunda Islands and the Philippines, and is very common in the vast coastal mangrove swamps. There are several subspecies which vary in colour. This snake stays always in the tops of the trees or tall shrubs, where at night it catches lizards, frogs and birds.

KING COBRA or HAMADRYAD
Ophiophagus hannah

The King Cobra, or Hamadryad, is found in eastern India, the whole of south-eastern Asia, Indonesia and the Philippines. It is the largest of the venomous snakes, and reaches a length of up to 5.5 m. It varies in colour, being yellowish, brown, olive-tinged or black, with different numbers of stripes, often on just the front part of the body. The venom of the King Cobra is highly toxic, and as a result of its size, the snake can inject large doses. Anyone affected by the poison dies within a few hours unless a serum is administered, and neck or head bites can prove fatal within a few minutes. Even elephants have been killed by cobras, for an elephant can be bitten in the soft skin at the tip of its trunk. The King Cobra is nocturnal. It usually lives in sparse forests and jungles where there is dense cover of bamboo. In the breeding season it builds a large nest in a heap of bamboo shoots and leaves. The female brings the nest material in a coil at the front of her body and heaps it up until it is up to 50 cm tall and measures 1 m across. The nest has two layers. The bottom one serves as a chamber

for the eggs, while the coiled female guards her clutch in the upper part. The male has sometimes been seen to take over guard duties from the female. She lays 30—40 eggs with soft, leathery shells. The young hatch after 3 months, and scatter in the surrounding countryside, to fend for themselves. Their venom glands are effective immediately after hatching. Although there are many tales to the contrary, the King Cobra is not an aggressive snake. It usually tries to scare away an attacker or intruder by raising the front part of its body and spreading its hood. When it is disturbed on the nest it charges the offender but never goes in pursuit of it. Native snake hunters often seek out the nests of these snakes, killing the adults before taking the eggs, which they eat as a delicacy. After removing the head with its dangerous venom, they also cook the cobra meat. Cobras are always caught in the daytime, for then their eyesight is poor. In Burma the King Cobra is used by snake charmers, who in these regions are always females, the art of snake charming being passed from mother to daughter. The women catch new cobras each year, releasing the old ones in the wild. They do not take out the venom teeth. Burmese snake charmers do not use flutes like their Indian counterparts, but encourage the snake to perform by means of movements of the hands. It is an amazing spectacle, for the King Cobra can raise its body up to a height of 130 cm.
The King Cobra feeds on other species of snakes, both venomous and non-venomous, and on lizards.

MALAYSIAN MOCCASIN
Agkistrodon rhodostoma

The Malaysian Moccasin reaches a length of 80 cm, and is found in Java, Malaysia, and southern Viet-

nam. It is variable in coloration, its ground colour being reddish, greyish or fawn, while its abdomen is either uniformly yellow or spotted green and brown.This moccasin is very common in some parts of Java, where it frequents plantations. It is feared by Europeans who are under the impression that its bite results in death within five minutes. However, natives working on plantations do not consider the moccasin's bite to be lethal, although its poison is strong. For all its high occurrence, moccasin bites are rare, for during the day the snake is not aggressive, and only bites when it is inadvertently stepped upon, and at night, when it becomes active and hunts, there is nobody working on the plantations. The Malaysian Moccasin feeds on frogs and small rodents. Unlike other species of moccasins which bear live young, the female of this species lays 12—30 eggs in a depression which she digs out herself.

INDIAN BAMBOO VIPER
Trimeresurus gramineus

The Indian Bamboo Viper reaches a length of only about 1 m. This green tree snake is an inhabitant of south-eastern Asia, where it is known as the 'banana snake', because it is almost always found in banana

Indian Bamboo Viper

trees. It is greatly feared both by the native population and by foreigners, who believe that it is very dangerous and that it slides on to the heads of people walking under the trees on which the snake lives. They maintain that the viper immediately bites its victim's neck, killing him within a few seconds. It is true that this snake sometimes falls on someone's head. The young snakes in particular are not expert at holding onto branches. However, the rest is quite untrue, bites being very rare and resulting at most in some degree of swelling. This viper is strictly arboreal, hunting smallish lizards, frogs, or even the young of small mammals in the trees. A single tree can be home to several vipers, particularly in the breeding season. The female bears 6—12 live young, which

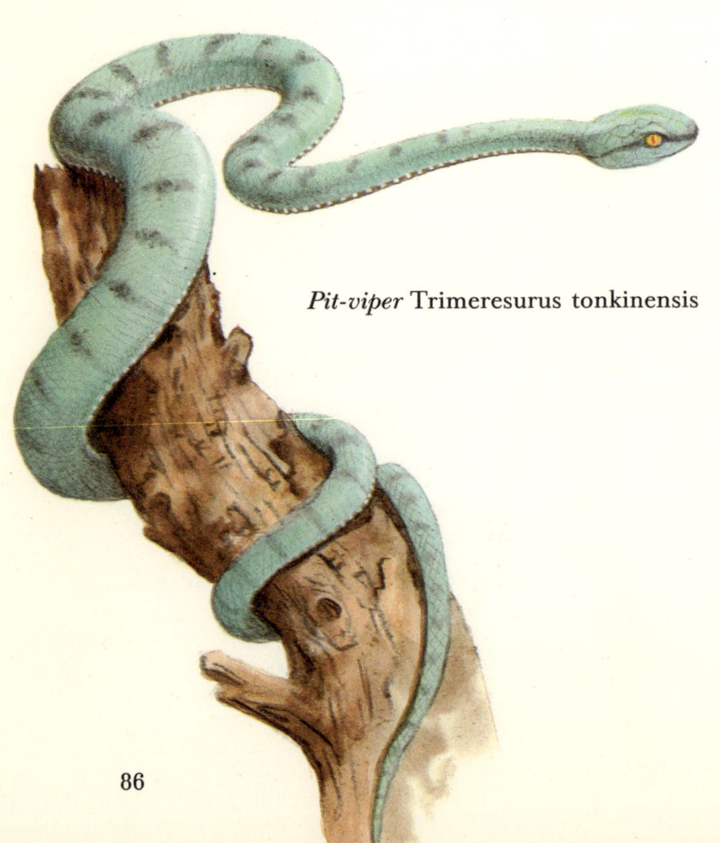

Pit-viper Trimeresurus tonkinensis

grip the branches with their weak prehensile tails, often falling off when a man or an animal brushes against the tree. It is this which has given rise to the superstition that the 'banana snake' is aggressive.

PIT-VIPER
Trimeresurus tonkinensis

The pit-viper *Trimeresurus tonkinensis* is distributed in northern regions of Vietnam. This arboreal snake reaches a length of 80 cm. Although it is abundant in some areas, it is rarely seen because its greenish or bluish-green colour blends with the foliage of the evergreen trees. During the day the snake rests, coiled around a twig, and hidden among the leaves. It comes out to hunt after nightfall, climbing along the branches and seeking small geckos and frogs. The female gives birth to about 10 live young.

MALAYAN FLYING FROG
Rhacophorus reinwardtii

The Malayan Flying Frog is found in the forests of Sumatra and Java. Its toes are connected by broad webbing, enabling the frog to make leaps up to 15 m long. Moreover, when it jumps, the frog flattens its body, draws in its abdomen and spreads its webbed toes wide, so that the air supports it as it glides. It cannot, of course, actually fly. When it lands on a leaf or branch, it grips, using the suckers on the broad ends of its toes. The diet of this frog is composed of insects, larvae and spiders.

JAPANESE GLIDING FROG
Rhacophorus schlegelii

The Japanese Gliding Frog is native to Japan, where it lives in trees and bushes in forests and thick scrubland. It can cover a distance of several metres by leaping in the same way as the previous species. It catches insects and other invertebrates. The female lays her eggs in a damp hole in which she has placed a special foam. She uses her hind legs to beat up the foam which protects the eggs and keeps the tadpoles moist as they develop. Although this usually takes place near water, the tadpoles of this species can survive without water.

Japanese Gliding Frog

Malayan Flying Frog

SPIDER
Myrmarachne plataleoides

The spider *Myrmarachne plataleoides* lives in and along the edges of thin woodland in Sri Lanka. It is a remarkable species, about 15 mm long, which perfectly imitates ants. It uses only three pairs of legs for jumping, stretching the front pair forward towards the ground like antennae. It does not spin webs but seizes small insects such as ants by pouncing on them. It lives always near colonies of the Red Weaver Ant (*Oecophylla smaragdina*).

STICK INSECT
Cyphocrania gigas

The stick insect *Cyphocrania gigas,* one of the largest species of its genus, reaches a length of about 25 cm. It inhabits the tropical zone of southern Asia, and is found in bushes and trees in thin woodland. It resembles a dry twig and this enables it to escape the attention of predators, although it is often captured by large insectivorous birds or lizards. This insect hatches from an egg and sheds its skin several times during its development, before reaching its final size. It has well-developed mandibles and feeds mainly on the leaves of various bushes. Several hundred related species of all sizes are found throughout tropical Asia.

MOVING LEAF
Phyllium siccifolium

The Moving Leaf is a leaf insect of south-east Asia and the adjacent islands. Its flattened legs and very broad, flat body make it look exactly like a green leaf.

Spider Myrmarachne plataleoides

Stick insect Cyphocrania gigas

Moving Leaf

Mantis Theopompa serrillei

Cicada Tosena seebohmi

standing its camouflage, it often itself falls prey to insectivorous birds and lizards.

The female lays her eggs in special foam cases which are attached to plants. The foam soon hardens, protecting the eggs from predators and from the effects of the climate. The newly hatched larvae feed mainly on aphids and other small insects. They shed their skins several times during their development, each time becoming more like the adults in appearance.

There are many other species of mantis in tropical Asia. Some have bizarre spines on their legs and bodies. Some species reach a length of 15 cm, while others resemble flowers, attracting insects, but at the same time avoiding insectivorous predators.

CICADA
Tosena seebohmi

The cicada *Tosena seebohmi* is one of the largest species of its genus, reaching a length of 7.5 cm and having a wingspan of over 15 cm. It is native to south-eastern Asia, but it also occurs in Australia. It is found in tall bushes and trees, where it sucks sap from the twigs. The female makes incisions in thin shoots and lays her eggs in the holes. The larvae hatch after 5—7 days, crawl down to the ground, and bury themselves in the soil where they suck the sap from the roots of the trees and bushes. After several years, when they become fully grown, they climb back up into a tree and attach themselves to a branch. Here their skins burst and adult insects emerge.

Although the males make a loud, distinctive sound, they can rarely be discerned among the greenery. Cicadas are extremely wary and become silent or fly away the moment they are approached. The well-known sounds are made by two V-shaped muscles at the base of the abdomen which cause a special plate to vibrate and so produce loud sounds. In many places the natives catch cicadas and keep them in tiny cages for entertainment. In ancient China, cicadas were symbols of youth, strength and life.

Many other species of cicada of all sizes are found throughout tropical Asia.

Indeed, many natives believe that these insects originated from leaves that grew legs. The female is about 10 cm long, while the male is smaller and has shorter wing-cases. In contrast, he has longer antennae. Leaf insects are able to fly, but they seldom take to the air. The large wing-cases of the female are a perfect imitation of leaves including the vein structure. Leaf insects usually sit motionless and invisible among the leaves on which they feed at night. When they are disturbed, they stand up on their thin legs and begin to swing. This often deters smaller insectivorous birds. The female lays large greyish eggs, which resemble seeds. These insects moult several times before they are fully grown.

MANTIS
Theopompa serrillei

The mantis *Theopompa serrillei* is distributed throughout south-eastern Asia and the Greater Sunda Islands. It has a broad, brown-mottled body, about 5 cm long. Its protective shape and coloration enable it to blend with the background in the branches of trees and bushes. Though it has wings it can only fly for a short distance. It is well-known for its predatory behaviour. It lies in wait for hours until its prey approaches, then it seizes the insect, gripping it with the spines on its front legs, while it eats the soft fleshy parts. Notwith-

COMMON MAP BUTTERFLY
Cyrestis thyodamas

The Common Map Butterfly ranges from India to

New Guinea. This large species has a wingspan averaging 65 mm. It rests with spread wings on the ground or on the bushes, choosing half-shaded places in the jungle. Here the map-like pattern on its wings makes it almost invisible. However, in flight it is frequently caught by insectivorous birds. This butterfly is often seen drinking from the edges of puddles or from dew-drops.

INDIAN LEAF BUTTERFLY
Kallima inachus

The Indian Leaf Butterfly inhabits forested areas of India and Sri Lanka. It has a wingspan of 85 mm, and, unusually for butterflies, it has projections on the tips of its forewings. These formations are of great importance. When the butterfly rests on a twig and folds its wings, the underside looks just like a large leaf, complete with central vein and stalk. This enables the butterfly to escape attention among the dense branches. When the butterfly is disturbed it quickly spreads its wings, revealing two eye-like spots which often scare away such predators as small birds or lizards. Butterfly hunters lure these butterflies by means of overripe fruit, the scent of which attracts them from far afield.

Other species of this genus are also found in southeast Asia.

MOTH
Erasmia sanguiflua

The moth *Erasmia sanguiflua* is one of the largest and most colourful species of its genus. It has a wingspan of about 95 mm. During the day, it flies, apparently hesitantly, over meadows or along the edges of forests, often alighting on tall plants. At night it shelters beneath large leaves. This beautiful moth is native to southern Asia.

Moth Erasmia sanguiflua

EDWARD'S ATLAS MOTH
Attacus edwardsii

Edward's Atlas Moth

Edward's Atlas Moth, with a wingspan of up to 18 cm, is one of the world's most beautiful moths. It is found in the mountain ranges of India, being most abundant in the foothills of the southern Himalayas. During the

day it rests on the branches of trees, and after dark it flutters along the edges of forests or in gardens. The female lays her eggs on the leaves of deciduous trees. The caterpillar is predominantly green and its body is covered with blunt outgrowths. It reaches a length of as much as 10 cm before pupating.

SWALLOWTAIL
Papilio hectorides

The swallowtail *Papilio hectorides* has a wingspan of about 10 cm, and is one of the most colourful species of butterfly. It is found throughout India, along the edges of forests, in plantations, or in parks and gardens. It glides in flight but moves relatively quickly. In the early morning these swallowtails fly down to drink from small pools. They feed on the sweet juice of overripe fruit.

RAJAH BROOKE BIRDWING
Trogonoptera brookianus

The Rajah Brooke Birdwing is a beautiful butterfly with a wingspan of up to 14 cm. The male is yellow, green or blue, while the female is plain chestnut or blackish in colour. This butterfly inhabits the forests of Borneo and Sumatra, where it flies high above the treetops, often gliding for long periods without moving its wings. As it seeks nectar from flowers blooming high above the ground, this butterfly is difficult to capture. However, it sometimes visits native huts in the heart of the primary forests, attracted by the scent of discarded fruit. Entomologists use such fruit as bait to catch these swallowtails. The caterpillar can project a colourful fork from the back of its neck in the same way as do the caterpillars of all swallowtail butterflies. The fork is effective in scaring away predators.

Rajah Brooke Birdwing

Owl Moth

Indian Moon-moth

OWL MOTH
Brahmaea wallichii

The Owl Moth has a characteristic, highly intricate
and dense wing pattern, and its wingspan is in excess
of 15 cm. It occurs in forested areas of tropical Asia,
flying among the trees by night, and resting on them
by day. The caterpillar has a bizarre appearance. It is
yellowish-green and is patterned in black and yellow.
In its first larval stages it has 4 outgrowths on the
front part of the body and two on the hind part. These
spines look like pieces of twisted wire and are shed
during the last moult. When the caterpillar is disturbed, it squirms vigorously, vibrating its spines and
often deterring smaller predators. Its entire development lasts only five weeks. Before pupation, the caterpillar turns a rich orange colour. It then crawls to the
ground and pupates in soft soil.

INDIAN MOON-MOTH
Actias selene

The Indian Moon-moth has a wingspan of up to
12 cm. It is distributed in forested areas from Japan to
southern China, India and Sri Lanka. During the day
it hides away among the large, dense leaves of trees
and bushes, and at night it flies along the edges of
jungles or in thin woodland. The female lays her eggs
on the leaves of various trees, frequently walnuts. The
caterpillar is greenish in colour, with reddish tuber-
cles.

LARGE CARPENTER BEE
Xylocopa caerulea

The large carpenter bee *Xylocopa caerulea,* which is
about 1.5 cm long, inhabits the forests of Sumatra.
The female of this species chews out a vertical nesting
tube 2 cm across and 30 cm long, in a post or a dead
trunk. She places a large quantity of pollen in the bottom of the tunnel, combining it with honey before laying an egg on the mixture. The bee then constructs
a partition made of wood debris and saliva, to close
off the chamber. Then she constructs another chamber on the roof of the first one and continues in this
way until the entire tunnel is built up. The larvae
hatch within a few days and feed on the prepared mixture for 3 weeks, by which time they are fully grown.

Large carpenter bee Xylocopa caerulea

Violin Beetle

Ground beetle Coptolabrus coelestis

After pupation, the adult bees bite their way out
through the wood, if the wall is weak, or alternatively
wait for the bees in the higher storeys to make way.
The females seek tubes vacated by other insects, in
order to build their nests. Large carpenter bees are
sometimes found in the beams of old houses.

VIOLIN BEETLE
Mormolyce phyllodes

The Violin Beetle, which reaches a length of about
85 mm, is found in the forests of Sumatra and Java. It
has a very flat body, well-adapted for living under the
bark of fallen trees. It is nocturnal, coming out after
nightfall to prey particularly on earthworms, but also

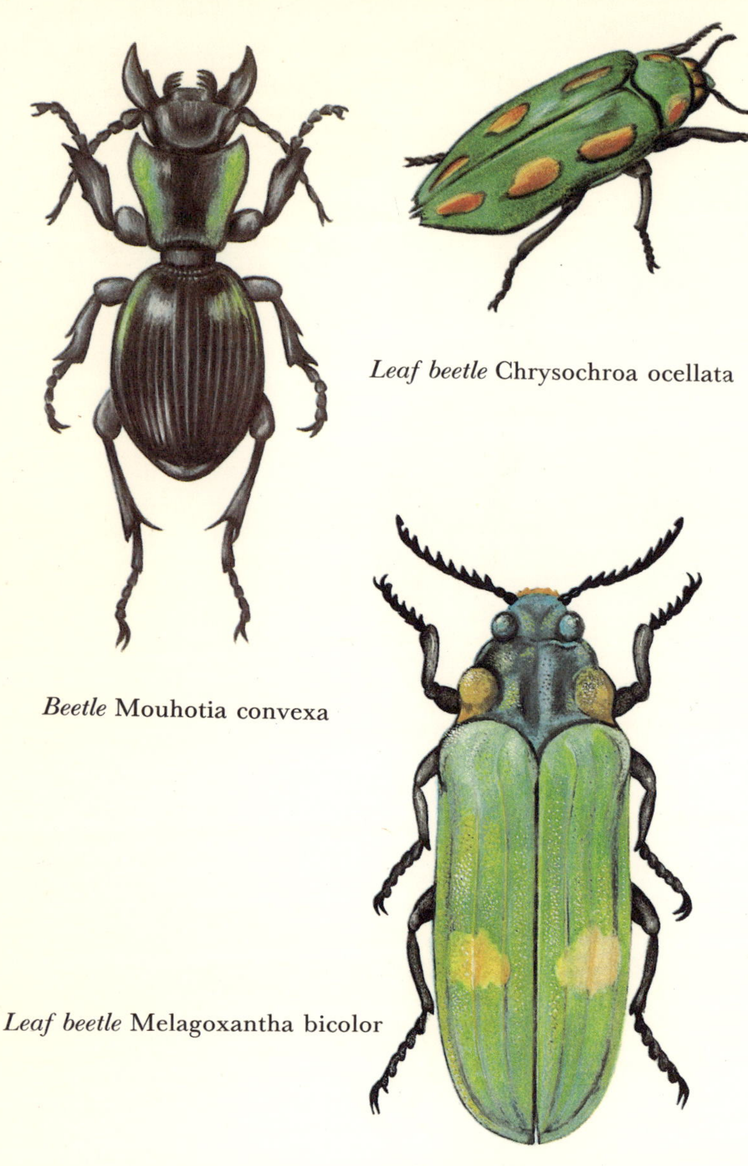

on various insects and their larvae. When it is in danger, it exudes a foul-smelling liquid that paralyzes small creatures which attack it. The female lays her eggs in holes beneath fallen logs. The predatory larvae pursue small invertebrates.

Leaf beetle Chrysochroa ocellata

Beetle Mouhotia convexa

Leaf beetle Melagoxantha bicolor

GROUND BEETLE
Coptolabrus coelestis

The ground beetle *Coptolabrus coelestis* is one of the most handsome species of beetle. It reaches a length of about 40 mm, and has wing-cases with a beautiful metallic sheen. This beetle is found in warm regions of eastern Asia, being most abundant in subtropical China. It lives in forests, in damp places beneath flat stones or fallen trees. It hides during the day, coming out in the dark to feed on worms, insects and insect larvae. It eats the soft parts of its prey, chewing them and mixing them with an intestinal secretion which quickly decomposes the flesh. The beetle then sucks in the liquid food. The female lays her eggs in holes in the ground.

BEETLE
Mouhotia convexa

The beetle *Mouhotia convexa,* which is about 40 mm long, inhabits south-eastern Asia. It has a massive body and extremely powerful mandibles, which scare away many creatures. Although this small beetle is not dangerous to larger animals, it can inflict painful bites. It lives in woodland, beneath stones or fallen trees, and hunts at night, catching worms, insects and other small invertebrates.

Stag beetle Odontolabis delesserti

Rhinoceros Beetle

LEAF BEETLE
Chrysochroa ocellata

The leaf beetle *Chrysochroa ocellata* is native to southern Asia. It is about 40 mm long. It is an inhabitant of forests and plantations containing large trees. It takes to the wing on sunny days, often settling on dead tree trunks. Here the female lays her eggs in the wood. The larvae take 2—3 years to develop and subsequently pupate in a small chamber. When the adult leaf beetle emerges it bores its way out of the trunk.

LEAF BEETLE
Melagoxantha bicolor

The leaf beetle *Melagoxantha bicolor,* a beautiful shining blue or green beetle, up to 75 mm long, is found in India. It lives in forests where it flies among the trees on sunny days. The females seek out cacao trees, in whose trunks they lay their eggs. The larvae live in the wood, and reach a length of up to 15 cm before they pupate. When these larvae are present in large numbers, much damage can be caused to the timber.

STAG BEETLE
Odontolabis delesserti

The stag beetle *Odontolabis delesserti* is found in the deciduous forests of south-east Asia. The female measures only 5 cm, but the male is up to 10 cm long, and has huge branched mandibles. In flight he resembles a miniature aircraft. He uses his mandibles in fights with other males. During the day, stag beetles move along tree trunks, seeking places where the bark is broken and sap is flowing out. The female lays her eggs in the rotting trunks of dead trees or in stumps. The larvae remain there for 3 years and pupate when they reach a length of about 15 cm. The male mandibles can be distinguished even in the pupal stage.

RHINOCEROS BEETLE
Chalcosoma atlas

The Rhinoceros Beetle can grow to a length of 13 cm. It has three outgrowths on its carapace and one upward-turned, dented projection on its head. The fe-

Tropical Leaf Beetle

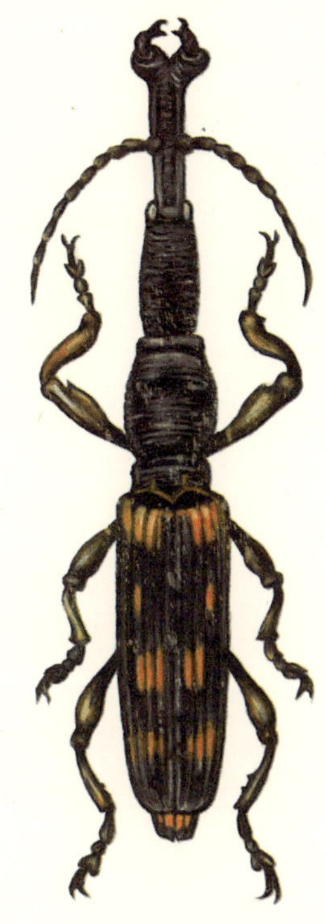

Beetle Butrachelus temmincki

male is considerably smaller and lacks these strange formations. This beetle inhabits the vast primary forests of the Greater Sunda Islands, being most abundant in Java and Sumatra. It also occurs in Sulawesi.

TROPICAL LEAF BEETLE
Sagra buqueti

The Tropical Leaf Beetle reaches a length of 3 cm and has enormous hind legs, though these are more slender in females. This insect lives in the trees and bushes of Java, being quite common in some localities. It feeds on leaves and green shoots.

BEETLE
Butrachelus temmincki

The beetle *Butrachelus temmincki* is the largest of the family of Long Beetles. This bizarre beetle reaches a length of up to 8 cm. It has an elongated body and its head is as long as the body itself. The antennae are situated at the tip of the head. This beetle lives on woodland flowers or beneath bark, where it seeks small insects and their larvae.

MOUNTAINS
AND
ROCKY TERRAIN

While northern Asia is covered with vast lowlands, the southern border of the central area of the continent is in contrast dominated by the highest and most massive mountain ranges in the world. In the west the colossal Caucasus soars between the Black and Caspian Seas, and in the south-east the bare Iranian Plateau continues east to the Hindu Kush. This range is subject to extremely severe climatic conditions, especially in winter, and many settlements in the area have been totally obliterated by avalanches. Further north the Pamirs, The Roof of the World, rise, and in the northernmost area, ranging east from Tashkent, is the Tien Shan massif. The Altai Mountains, with peaks exceeding 4 000 m in height, tower over the countryside of Central Asia. To the east of the Pamirs extend the mountain system of Karakoram and Tibet, and to the south this gigantic massif culminates in the soaring Himalayas, containing the highest peak in the world, Mount Everest, in Tibet, which reaches a height of 8 847 m. Many courageous climbers have lost their lives attempting to reach its snow-covered summit, but scores of mountaineers have conquered this giant among mountains. The Himalayan system extends south-eastwards to the Indonesian islands, though here its pinnacles are much lower. The Indonesian islands are characterized by the many volcanoes which rumble ominously and betray their activity with palls of grey smoke over their craters.

Nowhere in the world can one find a climate so capricious as that of the Himalayan region. The mountain ridges and peaks above 6 000 m are permanently covered by snow and huge glaciers, many kilometres long, feed mountain streams with crystal-clear water all year long. While from the freezing ridges gales thrust death-dealing avalanches into the valleys, low-lying areas enjoy subtropical weather, and south-facing slopes are covered by evergreen forests and humid tropical rainforests. Here, among the lush tangle of lianas and bizarrely-shaped tree trunks, grow rich clusters of magnificent pink and white orchids.

As well as possessing the highest peak in the world, the Himalayas are the source of the world's highest river, for at a height of 5 400 m the Dud Kosi River flows from the Kumbu Glacier in Nepal. In summer, the glacier forms a beautiful lake about 200 m below the source. From the lake, blue icy water rushes in a shallow, stony bed, pushing its way through rocky clefts and mountain passes and cascading down the valleys, until after 80 km, having fallen through a height of 4 km, it settles in a wide, massive stream.

The Himalayas are also home to the plant which grows at the highest altitude of all plants in the world. The chickweed *Stellaria decumbens,* with bell-shaped flowers, forms low, fragile clumps among the stones. It grows at a height of 6 130 m, in a hostile environment, but even its flowers are able to withstand the icy winds and the snowstorms. Other plants also survive at incredible heights. *Delphinium brunonianum,* with its large, dark violet flowers, is a typical alpine species found at heights from 3 600 to 5 400 m. *Meconopsis horridula* of the poppy family, which is found from 3 900 to 5 100 m, is one of the most beautiful of the Himalayan plants. It has large, sky-blue, bell-shaped flowers and leaves with long yellow thorns. The lion's foot *Leontopodium stracheyi* is found in the Himalayas and in other mountain systems of this area, while a related species, *Leontopodium ochroleucum,* also abounds here and ranges across China to the Japanese mountains. During the short spring season, a variety of gentians bloom on the Asiatic mountain slopes. Carpets of large, violet-blue flower of the species *Gentiana falcata* cover the ground beside the pale blue shining bells of the species *Gentiana depressa.* Higher up grow loaf-shaped cushions of low dense shrubs with tough needle-like leaves and tiny, inconspicuous pinkish flowers. This is *Acantholimon lycopodioides,* a common plant of rocky slopes. The cress *Cardamine loxostemoides* is a characteristic spring plant. It adorns the mountain meadows with its pink flowers. So also do the large, yellow-centred, pinkish rosettes of *Aster flaccidus.* Himalayan Balsam *(Impatiens glandulifera)* is another montane species. It opens its pinkish flowers in spring, and is able to shoot its ripe seeds as far as 3 metres from the parent plant. The pink primrose *Primula moorcroftiana* is found at heights up to 4 000 m, beneath glaciers and among broken stones and tufts of tough grass, and heads of the cestusroot *Saussurea gnaphalodes* emerge timidly among the stones. In June, at heights of 3 600—4 500 m, thousands of anemones and violet blue irises cover the mountain meadows. In spring, the entire area shines with large, fiery red poppies of the genus *Papaver,* although snow-white and orange-coloured forms are also found in the Himalayas. Roses bloom up to a height of 4 500 m on grassy slopes, and the Musk Rose *(Rosa moschata)* abounds in central south-western Asia.

The most attractive plants of the Asiatic mountains are rhododendrons. Their extravagant beauty enlivens the landscape not only in the zone of alpine woodland above a height of 3 000 m, but also far above the forest line, up to an altitude of about 5 000 m. In Nepal, rhododendrons grow up to 20 m tall, and smaller species range to Malaysia and as far as southern Japan. These extensive areas of rhododendron create a splendidly colourful mosaic on the rocky slopes, and at the same time provide safe shelter for a variety of alpine fauna.

Typical of Asian mountain flora are junipers of the genus *Juniperus,* growing at heights up to 5 000 m. There are scores of species, ranging from giants 20 m tall, to stunted shrubs which grow at the highest altitudes on sheltered slopes. Lower down in the zone of the upper forest line above 3 000 m, junipers alone often cover extensive areas of the mountain slopes. Prickly juniper thickets are ideal haunts for many species of pheasant, as well as for other birds and mammals.

The forest line varies from area to area, occurring at a height of about 2 400 m in the Caucasus, and at 3 600 m above sea level in tropical and subtropical localities. Asiatic forests contain many species of both coniferous and deciduous trees. Some conifers, such as the Long-leaved Indian Pine *(Pinus longifolia),* their trunks and boughs deformed by the strong winds, grow both in ravines and on mountain slopes. In many places, conifers form continuous areas of thin woodland, only a few metres tall. At night they shelter a variety of diurnal birds in their crowns, especially pheasants, raptors and birds of the raven family. Firs of the genus *Abies* form characteristic vast woods with dense rhododendron undergrowth, reaching to the forest line. Oaks of the genus *Quercus,* such as *Quercus incana,* whose twisted trunks grow to a height of only a few metres, abound in this area, while on lower slopes the oak forests are denser and the trees are more massive. In the western Himalayas, oak forests are found up to a height of 3 000 m. The Deodar Cedar *(Cedrus deodara)* has a conical trunk and dark bluish-green needles, up to 5.5 cm long. Barberry shrubs of the genus *Berberis* are found at heights of over 2 000 m in the mountainous areas of Asia. *Berberis chinensis* and *Berberis brachypoda,* whose yellow, scented flowers lure hundreds of insects, or *Berberis julianae,* with its reddish berries and thorns up to 3 cm long, are found in the mountains of China. Shrub-like trees of the genus *Ficus,* whose large, dark green leaves provide shelter for many varieties of small birds, grow in the mountain passes up to a height of over 2 000 m.

Birch trees are also common in mountainous areas, particularly from the Altai Mountains eastwards. They occur mostly in stunted forms, at heights in excess of 2 500 m. Stunted willow trees, and various species of spruce of the genus *Picea* form extensive woods in many localities. The Oriental Spruce *(Picea orientalis)* of the mountains of Asia Minor and the Caucasus, is conical in shape, with branches growing low down on the trunk. It is covered with short needles, 6—10 mm long. In contrast, the West Himalayan Spruce *(Picea morinda)* has needles up to 40 mm long. Over 10 species of spruce grow in the mountains of central China, and spruce such as the Sakhalin Spruce *(Picea glehnii)* can be seen on the mountains of Japan.

Another important mountain plant is bamboo, which grows in tropical and subtropical areas at heights up to 4 000 m. This grass forms extensive thickets in some areas, growing less tall with increasing altitude.

Hundreds of other plant species can be found on the mountains of Asia, including grasses, herbs, shrubs and trees.

Both the mountains and the extensive rock-covered areas have characteristic representatives in the animal realm. Many of them are endemic, that is to say, confined to specific localities. These species include the Giant Panda and the Lesser Panda, the former being one of the rarest animals in the world.

The Asiatic mountains rank first not only in the numbers of plants growing at extreme altitude, but also in the numbers of animals flourishing at great height. Small pikas, or piping hares, relatives of true hares, live up to a height of 6 000 m in the Himalayas. Beautiful butterflies fly over the mountain meadows of the Hindu Kush, the Pamirs and the Himalayas, at heights of 4 000—5 000 m. They include colourful species such as *Aglais cashmirensis, Polyommatus sarta* and *Colias erate,* and most frequent of all, the Apollo butterfly *Parnassius acco,* found up to a height of 5 800 m. These are only a few of the many butterflies of the mountains, and the trees and shrubs of the mountain slopes are also inhabited by a variety of beautiful beetles.

One of the largest Asiatic mammals is found at heights from 4 500—6 000 m. Here, where the temperature in winter drops below 40° C, and snowstorms alternate with icy, blasting winds, the wild Yak is able to withstand the bitter cold, and the domestic form has been bred for thousands of years. Yaks provide milk with a high fat content, they are used as beasts of burden, and their dried dung is used as fuel.

The mountain slopes and bush-covered plateaux are home to a variety of fowl-like birds, pigeons and raptors. Eagles of many species cruise high above the rocky passes and valleys, and over alpine steppes covered with wormwood, as they scour the ground for prey. Not even the speedy Long-tailed Marmot *(Marmota caudata)* or Royle's High-mountain Vole *(Alticola roylei)* escape the eagles' keen eyesight. Lammergeiers and vultures search the countryside hour after hour for the carcass of a deer, a yak or a wild sheep, in order to feed on the flesh. The mountains are also the home of large carnivores such as the Asiatic Black Bear and the Snow Leopard, and even the Spotted Leopard sometimes wanders above a height of 4 000 m. Bears and leopards often visit the outskirts of settlements to take domestic sheep or goats, or even some larger animal. They seldom attack men, except when they are wounded and are unable to escape.

To make this survey complete, one last creature must be mentioned. It reportedly lives in the Himalayas, Pamirs and Hindu Kush, and according to some it can be found even in the Tien Shan and the Altai Mountains. Yet this creature cannot be found in any zoo or wildlife park, and there are no records of it in any museum. Not one of the best photographers has taken its picture. All that exists are a few photographs of its tracks in the snow. What is this mysterious inhabitant of the high Asiatic mountains? It is no other than the abominable snowman, or yeti. This creature dominates its territory, and local people dare not raise their voices when mentioning it, for fear of turning its anger against them. The native Sherpas of Nepal appear to be quite familiar with this fearsome ruler of the mountains, and are convinced that it exists. The supposed existence of this creature is mentioned as early as 1857 by J. D. Hooker, who describes it in his book about the Himalayas. Although he never saw the snowman himself, people to whom he talked described it as a hirsute man inhabiting the inaccessible snows of the mountains. W. A. Weddell, in a book published in 1899, says he saw large tracks in the Himalayas, which the natives believed to be those of a hairy man, living in the region of eternal snow and ice. The author expressed some doubt, admitting that the tracks could have been left by a bear. In 1915 a forester reported that he had seen an erect, hairy, manlike animal, about

120 cm tall, high up in the mountains. Over the years many expeditions visited the Himalayas and other mountain ranges of Asia, and their members climbed the highest peaks including Mount Everest. Some participants 'met' the abominable snowman, whose footprints were photographed by Eric Shipton's expedition in 1951. The footprints were 30 cm long, and very broad, and could not have been made by any known animal. From the depth of the footprints, it was estimated that the animal had to weigh about 90 kg. Experts studied other footprints, and a variety of theories resulted. One proposed the existence of descendants of prehistorical apes of the genus *Gigantopithecus,* but this is a fantastic theory, based on insufficient knowledge of the facts. Another suggested that the tracks were of men of an Eskimo tribe, and yet another that the snowmen were people who had once sought refuge in the mountains and had survived there with their offspring. Some experts are of the opinion that the abominable snowman is a so far unclassified species of ape or bear.

Reports of the abominable snowman continue to multiply, books are published, articles appear in respected journals, and groups of mountaineers set off on special expeditions with the mission to unravel the mystery. In 1954 the *Daily Mail* financed an expedition to the Himalayas, which included a team of eminent zoologists and anthropologists. Not one of the experts saw a yeti. They photographed some footprints which local people claimed were made by the snowman, but after careful study, it transpired that the tracks could have belonged to a bear or to a snow leopard. Certainly some footprints were unusually large, but this could be accounted for by the snow having melted along the edges and frozen again at night. Himalayan expeditions have continued, but always with the same result. The local people talk, but no impartial outsider has yet seen an abominable snowman.

At one time, the existence of a collection of scalps, pieces of skin, and even mummified hands and other items testified to the existence of the yeti. These objects were kept, like holy relics, by the Himalayan monks. For a long time the monks were reluctant to lend them to the experts, but in the end they did. However, all the remains proved to be those of known local animals.

Nevertheless, many experts still believe that an unknown, so far undiscovered animal may live in the vast, almost inaccessible wastes of the Asiatic high mountains.

Shrew Hedgehog

SHREW HEDGEHOG
Neotetracus sinensis

The Shrew Hedgehog inhabits the provinces of Szechwan and Yunnan in China, and it is also found in northern Burma, Thailand and Vietnam. It reaches a length of 23 cm, of which 8 cm is taken up by the tail, and it has long, dense, very fine fur. This species is confined to rainforests in mountainous areas, at heights from 2 100—2 800 m. It excavates burrows

Douc Langur

under stones, lining them with moss. It begins to hunt after nightfall, scurrying over the ground in search of earthworms, molluscs and insects. After a gestation period of 5—6 weeks, the female bears 2—5 young in the burrow. They are born blind and gain their sight after 3 weeks. They are suckled for about 2 months but begin to take solid food when they are 40 days old. There are usually two litters a year.

DOUC LANGUR
Pygathrix nemaeus

The Douc Langur is native to Laos, Vietnam and the island of Hainan. It measures up to 150 cm, but a half of this is taken up by the tail. It lives in small family groups composed of an adult male, and females with their young. Old males are sometimes solitary, while juvenile males without families of their own merge in communities of up to 30. During the day, langurs roam over the hillsides in the moist forests, or wander among tall shrubs along the edges of forests, or beside streams in mountain valleys. They have no fixed territory, and the groups often meet during their forays. Such encounters usually take place without any signs of conflict, but on rare occasions, the leading males fight each other. Langurs have regular sites where they gather to spend the night. They choose rocky situations with caves on steep slopes, often above water. They leave their shelters after sunrise and set out to forage. Langurs eat young leaves, shoots, blossoms and various fruits, and occasionally take insects or molluscs. On hot days, they descend to the water to bathe in the shallows, for they are quite good swimmers. After a gestation period lasting about 168 days, the female gives birth to a single young, which immediately clings to its mother's abdominal fur. Sometimes it is also carried by the male. The young langur is suckled for 10—12 months, and it begins to take solid food at the age of 2—3 months. Females mature after 3.5 years, males after 4 years.

STUMP-TAILED MACAQUE
Macaca arctoides

The mountainous areas of south-eastern Asia are inhabited by the Stump-tailed Macaque. This monkey is about 65 cm long and has either a red or a red-spotted face, and a short tail. It is able to withstand the

cold, and lives at heights up to 4 000 m, mainly on wooded rocky slopes on the banks of mountain streams. Here it wanders in groups of 20—40, or even sometimes of 100. Each group, led by the strongest male, roams by day through the bushes and low trees of its territory, foraging for shoots, leaves, fruit, berries, insects and small vertebrates. Sometimes these macaques take birds' eggs, and they often raid fields and plantations, causing considerable damage to maize and bananas. Stump-tailed macaques spend the night in caves, usually among rocks above a river. Following a gestation period of 170—180 days, the female bears a single young. The baby monkey clings to the fur on its mother's abdomen with its hands and feet, and is carried everywhere she goes. The young macaque is suckled for 14—18 months but begins to take regular solid food after 2 months. At the age of 180 days, young macaques are able to fend for themselves.

ENTELLUS LANGUR
Presbytis entellus

The Entellus Langur is one of the best-known monkeys, being sacred in India, where it has been protected since ancient times. As a result, groups of langurs often live in towns, on the walls of the temples. They raid gardens with impunity, and sometimes steal fruit from street traders, who are honoured by the visit of the sacred monkeys. The Entellus Langur sometimes measures over 170 cm including the tail, which is over 100 cm long. There are several subspecies differing in coloration and length of hair. This monkey is distributed from Pakistan, across Nepal and Assam, to India and Sri Lanka. In the mountains, it ascends up to a height of 2 500 m, but it comes down to the valleys in winter. During the day, it wanders around in groups composed of a leading male and several females with their young. Solitary males, though chased away by the leader, tend to stay nearby. The groups sometimes combine into troops of over 120. The Entellus Langur is indigenous to wooded slopes and bush-covered rocky sites, where it feeds on leaves, flowers, fruit, and occasionally on insects. Groups of monkeys sometimes gather in a vegetable field, spoiling the crops, but never being pursued. They are skilful climbers, and on the ground they can leap distances of up to 10 metres. Families spend the night in small caves or on high temple walls. After a gestation

period of 168—196 days, the female bears a single young. The newborn monkey clutches its mother's abdomen and suckles for up to 12 months. It begins to nibble young shoots and soft ripe fruit from the age of 3 months.

105

Capped Langur

CAPPED LANGUR
Presbytis pileatus

The Capped Langur grows to a length of 130 cm, half of which is made up of the tail. It is an inhabitant of tropical rainforests ranging from Assam, across Burma to southern China. This monkey lives mainly in

jungles, in rock-covered localities, particularly in river valleys. It gathers in family groups, roaming by day throughout the treetops where it feeds on leaves, shoots and various fruits. Before nightfall, the group returns to a regular site, usually a steep rock wall containing small caves, in which the langurs sleep. For half an hour or so they sit on the branch of a tall tree, high above their rocky shelters, carefully scanning the surroundings. After making sure that there is no danger, the monkeys descend, one after another, and climb into the caves. Following a gestation period averaging 168 days, the female bears a single young which she carries with her, clinging to her abdomen.

GOLDEN SNUB-NOSED MONKEY
Rhinopithecus roxellanae

The Golden Snub-nosed Monkey is found in Tibet and south-western China. Its total length is up to 140 cm, half of which is taken up by the tail. This monkey is characterized by its extremely thick fur, 15—18 cm long, which protects it from the cold of the high mountains. The Snub-nosed Monkey lives in thinly wooded areas at a height of over 4 000 m, in places which are covered by snow for most of the year. It lives in family groups, roaming the tree-covered slopes in search of leaves, shoots, fruits, birds' eggs and insects. It spends the night in caves in steep walls of rock among the bushes. After a gestation period of about 170 days, the female bears a single young, which she carries against her abdomen.

ASIATIC RED WOLF or DHOLE
Cuon alpinus

The Asiatic Red Wolf, or Dhole, is distributed from southern Siberia across China, central and southern Asia to Sumatra and Java. This carnivore has been exterminated in many areas and may well follow the same fate in India and Java, where it is mercilessly hunted. The Asiatic Red Wolf reaches a length of up to 150 cm including its 50 cm-long tail, and weighs 14—21 kg. It frequents forests in mountainous areas up to a height of 4 000 m. It usually forms packs of 5—20, but sometimes as many as 30 individuals may hunt together by both day and night, working together to surround larger prey such as young deer or wild sheep. Red wolves are not fast runners and never

Golden Snub-nosed Monkey

chase their prey for very long. They sometimes catch
fowl-like birds, insects, such as large grasshoppers,
and molluscs. When resting, they shelter in caves or
holes in river banks, only seldom digging their own
burrows. They are very quiet and inconspicuous in
their territories, but on scenting a tiger, which is their
chief enemy, they bark ferociously, warning fellow
members of the pack and other animals as well. In the
breeding season, the Asiatic Red Wolf lives in pairs,
but several females sometimes rear their offspring together. The female bears 2—9 young after a gestation
period of 60—63 days, usually in January or February. The young are born well-furred, but they only
gain full sight when they are 13—15 days old. By the
end of the first month they are able to nibble at meat
from their parents' catch, but they continue to be
suckled for a further 4 months.

ASIATIC BLACK BEAR or MOON BEAR
Selenarctos thibetanus

The Asiatic Black Bear, or Moon Bear, is widespread
from eastern Iran across India to the north of south-
eastern Asia, and extending to China, central and
eastern Siberia, Korea, Japan, Taiwan and Hainan. It
stands up to 80 cm high at the shoulders, is 180 cm
long and weighs over 120 kg. Its tail is only 10 cm
long, and there is a characteristic white mark shaped
like a crescent moon on its chest. It inhabits mountain
forests up to a height of 4 000 m, occurring in the
north in both taiga and tundra. In winter it does not
hibernate, but descends from higher situations to altitudes of about 1 500 m. It forages in the trees, being
a good climber, and it often spends the night among
dense treetops, where it feeds on eggs and nestlings,
wild bees' honey, insects, molluscs and other invertebrates. However, it concentrates mainly on plant
food, picking various fruits, eating leaves and shoots,
and occasionally taking growing crops from the fields.
It is known to attack larger domestic animals on occasion, and it is capable of killing a small cow or horse.
However, it may fall prey itself to the more robust
Brown Bear. The Asiatic Black Bear is normally active
during the day and at dusk, but in localities where it
is hunted, it comes out at night. For most of the year
it is solitary, staying with a mate only for a few days in
the breeding season, in June or July. After a gestation
period of 7—8 months, the female bears 1—2, or occasionally up to 4 cubs in a cave or a large hollow

Asiatic Red Wolf or Dhole

Asiatic Black Bear or Moon Bear

Lesser Panda

tree. The young are blind at birth, opening their eyes after 24—34 days. They leave the den when they are 48—60 days old and begin to feed independently after 85—90 days. Young bears are mature at the age of 2.5 years.

LESSER PANDA
Ailurus fulgens

The Lesser Panda is native to the high mountains from Nepal to northern Burma and south-western China. It reaches a total length of about 110 cm of which about 50 cm consists of the bushy tail, and it weighs 3—4.5 kg. This most beautiful carnivore is found at heights from 1 800—4 000 m, where it lives in forests and bamboo thickets. Old males usually

lead a secluded life, but families often stay together throughout the year. Pandas have their own territories with regular pathways, marked with excretions from the anal glands. They forage after nightfall, or even during the daytime on overcast days. They nibble at bamboo shoots, grass, roots and various fruits, and also feed on birds' eggs, the young of small species of birds, and insects. Pandas have regular places in which to rest and sleep, usually in a hollow tree, a cave or among dense branches.

After a gestation period averaging 130 days, the female gives birth to 1—2, or occasionally up to 4 young. These are well-furred at birth but only open their eyes after 21—30 days. They begin to take solid food after 16 weeks and mature at the age of 2—3 years.

Giant Panda

GIANT PANDA
Ailuropoda melanoleuca

In some ancient Chinese paintings a strange black-and-white bear is depicted, and though historians and zoologists knew these paintings well, for a long time they regarded the animal as purely imaginary. Japanese documents dating from October 22, 685, tell of the presentation by the Chinese Emperor to the Japanese Emperor of seventy bear skins and two live white bears. According to the documents, the bears were captured in central Asia. This no expert believed. Instead the bears were taken to be polar bears brought to China from the north — for how else could white bears be found in China? This theory was supported by the French Jesuit priest Armand David, who went to China as a missionary, and who was an enthusiastic naturalist and collector. Père David spent the winter of 1868—1869 at the missionary station in Cheng-Tu, capital of the distant western province of Szechwan. Here he heard about the beautiful white bear which lived in the mountains near the village of Moupin and which was worshipped by the local people. At first he thought this was just one of the many Chinese legends, but one day he decided to satisfy his curiosity, and he set out for Moupin. After some days of difficult walking and back-breaking climbing over the steep mountain paths, he reached Moupin on March 1, 1869. Here he set about his search for the legendary animal, but in vain. There was not even a similar kind of animal to be found there, so on March 11, he decided to turn back. Weary after his fruitless efforts, he stopped to rest at the home of a farmer named Li, who received him warmly and took him to his room. Exhausted, he sat down and then almost jumped out of his chair again, for on the wall in front of him was the skin of the white bear! In fact it was black and white, and Père David was overjoyed to realize that he had discovered the white bear of the old Chinese chronicles and of the villagers' tales. The bear was not a legend after all, but a real animal, and the Japanese Emperor had indeed received his gift from central China! Père David decided to stay and learn more from the local people about the strange creature. Their hunters offered to catch an animal for him, and left at once for the mountains returning on March 23 with a young bear. They had caught it alive but had killed it to make the journey back easier. The skin and skull were sent to Paris, where professor Alphonse Milne-Edwards recognized that this was a new species, not yet scientifically classified. From the teeth and paws he was able to show that the animal had nothing in common with true bears. The Giant Panda had been discovered.

The Giant Panda reaches a length of up to 1.5 m and a weight of 75—160 kg. It has only a very short tail. It is currently one of the rarest animals in the world, and the few zoos which have specimens have only 20 or so between them. In the wild it is restricted to the Hsifan Mountains in western Szechwan, where it lives at a height of about 2 000 m. It frequents vast bamboo thickets, for the leaves and shoots of this plant form the bulk of its diet, and it eats for 10—12 hours a day.

Occasionally it catches fishes in a small stream, or piping hares among the rocks. It is solitary, forming pairs only briefly in the breeding season. The female produces a single young which stays with her for 2—3 years. When it becomes independent, it roams the countryside in company with other juvenile pandas. Females mature after 4 years, while males often take as long as 10 years to reach sexual maturity. In winter, the Giant Panda hibernates in its snow-covered den.

SNOW LEOPARD or OUNCE
Uncia uncia

The Snow Leopard, or Ounce, is a resident of the high mountains ranging from eastern Turkestan to southeastern Tibet, extending in the north to the Aral Sea and Lake Baikal. It reaches a length of 2.5 m but over 1 m of this is taken up by the bushy tail. The weight of this handsome carnivore, covered in winter with long, thick fur, is 23—41 kg. The Snow Leopard is still abundant in some areas, but it is not easily observed because it is adept at concealing itself. It frequents mountain slopes, usually at heights from 3 000—4 000 m near the snow line in summer, while in winter it descends to lower land. It is solitary, each individual having an extensive territory with regular pathways and containing secluded places where it can bask in the sun and sleep. The Snow Leopard usually hides in caves or beneath overhanging rocks on steep slopes. From here it sets out to hunt, usually at dusk but often at night or even during the day. It lies in wait on a high vantage point, pouncing 6—15 m on to its prey. It mainly attacks wild sheep, goats, deer and wild hogs, but it also catches mice, marmots, piping hares and pheasants.

In the breeding season, the male joins his chosen mate for a few days. Before giving birth, the female lines her den with grass, lichen and fine hair plucked from her abdomen. In April—May, after a gestation period of 98—103 days, she bears 1—5 young. These are born blind, but open their eyes when they are 7—9 days old. They are suckled for 6 months, but from the age of 2 months, they nibble at prey brought by their mother. The playful leopard kittens leave the den after 45 days, but stay with their mother until the following winter. For this reason, there is usually only one litter every two years. Females mature after 2—3 years, males after 4—5 years.

Snow Leopard or Ounce

MUSK DEER
Moschus moschiferus

The Musk Deer inhabits coniferous forests and rhodo-dendron-covered slopes in mountain areas ranging from Nepal to China, Korea and eastern Siberia. It is found at heights of 2 600—3 600 m, in valleys with mountain streams. It stands about 60 cm high at the shoulder, is 100 cm long, has a tail measuring only 4 cm, and weighs 9—11 kg. Although it is classed among the deer, it has no antlers. Instead it has elon-

Musk Deer

gated upper canine teeth, developed as tusks, up to 7.5 cm long. These make effective weapons, being used with great dexterity. The Musk Deer lives in pairs or singly, defending a territory where it digs out shallow depressions in which to rest. It forages on regular paths, marked with a secretion from the anal gland, and feeds on leaves, grass, fallen fruit, moss and lichen. On sensing danger, such as a tiger or leopard, the deer makes a loud whistling call, warning all the animals in the neighbourhood. The rutting season takes place from November to January. After a gestation period lasting about 160 days, the female bears 1—2 spotted young in dense bushes. The calves are suckled for 4—5 months.

THOROLD'S DEER
Cervus albirostris

Thorold's Deer, a sturdily built species standing up to 150 cm high at the shoulder, is confined to the high mountains of Tibet and western China. The female is smaller, more delicately built and lacks antlers. In winter, the deer grow long, dense fur, protecting them from freezing winds. Adult males are normally solitary, but they form small herds in winter. The animals move along regular paths when foraging for food, their diet consisting of grass, leaves, fruits, bamboo shoots, and also moss and lichen in winter. In the rutting season, in September and October, the bulls fight to acquire a herd of cows. In July, or thereabouts, the female produces a single calf, or sometimes twins, in the undergrowth, after a gestation period of 220—240 days. The calf is suckled for up to 8 months but begins to feed on plant material when it is a month old. Thorold's Deer is preyed on principally by leopards, bears and wolves, the young being especially vulnerable.

WILD YAK
Bos mutus

The Wild Yak is found on the high plateaux of Tibet. It can be encountered up to a height of 6 100 m, but it descends to the shelter of lower valleys in winter, seeking protection from biting frost and severe snowstorms. It is a massive animal, up to 2 m high, weighing about 500 kg and having horns up to 1 m long. The cows are smaller and have less robust horns. The Wild Yak was domesticated some 3 000 years ago and has ever since been raised in many mountain regions of Central Asia. The cows yield only a small quantity of milk, but it is very nutritious and has a high fat content. The Wild Yak moves in groups made up of cows and calves. The bulls are often solitary, but in the rutting season they fight each other to obtain a herd of females. Yaks have their own territories which they roam in search of food. They consume grass, the leaves of bushes, bamboo shoots, lichen and moss. After a gestation period of 255—270 days, the female bears a single young, which often stands up half an hour after birth, and suckles when it is only an hour old.

Thorold's Deer

GORAL
Nemorhaedus goral

The Goral is widespread from the Himalayas to China, Korea and eastern Siberia. It reaches a height of 55—75 cm, a length of up to 130 cm including the tail, which is 20 cm long, and a weight of 22—32 kg. It dwells in rocky areas up to a height of 2 500 m, although it occurs in low-lying scrub in semideserts as

Wild Yak

Goral

TAKIN
Budorcas taxicolor

The Takin occurs in the high mountains east of Tibet, from where it extends to central China. Heavily built, it is 110—130 cm high at the shoulder and adult bulls weigh up to 275 kg. The massive horns are about 60 cm long. The Takin is found at heights of 2 000 to 5 000 m, where it frequents sparse mountain forests, rhododendron-covered slopes and bamboo jungles. It is usually active in the late afternoon and evening, but in winter it is active even during the day. It moves in small groups, but old bulls are solitary. Takins roam through their territories along narrow, well-trodden paths and they have regular places for rest and for defecation. They feed on grass, plants, leaves and fruits, and in winter they eat bamboo shoots in the shelter of the valleys. The female produces a single young, or occasionally twins, after a gestation period of 200—220 days. Before giving birth, she leaves the herd and hides away among the brushwood. Calves are born in March or April and follow their mother half an hour after they are born. When they are 14 days old, they are able to nibble juicy shoots and they are often weaned by the age of 2 months. Females mature after 2.5 years, males after 4—5 years.

MARKHOR
Capra falconeri

The Markhor is native to the mountain regions of Uzbekistan and Tadzikistan in the Soviet Union, and in Afghanistan, Kashmir and Pakistan. The male is 115 cm high, 140 cm long including the tail, which measures 15 cm, and he weighs about 120 kg. He has strong, spiral horns. The female is more slightly built, weighing 50 kg at the most. The Markhor is diurnal, but it is also active on clear moonlit nights. It climbs

well. It climbs steep, stony slopes with agility and speed, easily escaping its many predators. The males live in groups for most of the year, but old bulls are solitary. In the rutting season each male defends his own small territory, marked by means of glandular secretions which he spreads with his horns on to twigs or tall plants. Mating takes place between September and December. Following a gestation period of 210—240 days, and usually in May or June, the female bears a single, or sometimes two young in a secluded place in the undergrowth. The calf is able to stand up after only an hour. It is very active and follows its mother from the second day, staying with her for two years. Young gorals which have strayed from their mothers are often seized by large eagles. If the mother is near at hand when her calf is attacked she immediately charges at the raptor, using her pointed, dagger-like horns.

Takin

steep rocks with dexterity and leaps into ravines several metres deep, escaping from its predators through the narrowest of passes. The females and their offspring live in herds, and they are joined by the males in the rutting season. At this time many duels take place, and the loud crashing of horns can be heard far afield. However, the fights usually end without casualties. After a gestation period averaging 155 days, the female produces 1—2 extremely agile young. They are suckled for 5—6 months but stay with their mother for 2 years. The Markhor feeds on grass, plants, leaves and twigs, moss and lichen. On sensing danger, it makes a sharp, loud whistle.

WILD GOAT
Capra aegagrus

The Wild Goat is widely distributed from the southern Caucasus across Armenia to Pakistan, Iran and Turkey. There are several subspecies. The male is up to 105 cm tall and has massive horns, while the female is less robust, and has smaller horns. The Wild Goat is an ancestor of the domestic goat. It was domesticated in western Asia more than 9 000 years ago. The Wild Goat lives in large herds in rough mountainous areas, on rocky slopes covered with juniper, with various bushes or with deciduous forest. The herds are made up of females with their young and males up to 4 years of age. Old males are solitary or form independent groups outside the rutting season, which takes place from October to January. Fierce encounters take place for possession of the herds of females. After a gestation period lasting 155 days, the female bears 1—3 young. They are able to follow their mother from the second day, and skilfully climb over the rocks. They are suckled for 5 months. The Wild Goat feeds on leaves, grass, a variety of fruits, and on moss and lichen in winter. Its principal enemies are leopards and wolves.

TAHR
Hemitragus jemlahicus

The Tahr is distributed from south-western India to the Himalayas. It is characterized by its very long hair, which grows even thicker in winter. Tahrs vary in colour, ranging from fawn to blackish-brown. The male is about 100 cm tall and weighs 100 kg, while

Markhor

Wild Goat

Tahr

ARGALI
Ovis ammon

The Argali is the largest member of the sheep family. It is native to the Altai Mountains from where it extends to Lake Baikal and western Mongolia. The male reaches a height of 125 cm at the shoulder, a length of 200 cm, and a weight of up to 230 kg. The female is smaller and has more delicate horns. The Argali is found in mountainous areas, where it feeds on grass, flowers, young plants and the leaves of bushes. The rutting season takes place in October and November, and sometimes lasts until January. At this time the males often engage in combat. A single young, or sometimes twins, are produced after a gestation period of 150—190 days. The young follow their mother from the second day after birth. Females are mature at 1.5 years, while males reach maturity at 3.5 years.

the female is smaller and has more delicate, though equally long horns. The Tahr frequents tree-covered mountain slopes up to a height of 3 000 m, where it moves in groups of 10—40. While the herd is grazing, a few members stay on guard, to warn the others when danger approaches. Tahrs are extremely wary, taking flight across the steep slopes at the slightest suspicion. They feed on grass, all kinds of plants, and fruits. In the rutting season, in November and December, the males emit an intense characteristic odour. The female gives birth to one, or occasionally two young, in a secluded spot among the rocks, after a gestation period of 180—240 days. The young are very agile, and a few hours after birth they join the herd with their mother.

BAR-HEADED GOOSE
Anser indicus

The Bar-headed Goose, which is about 75 cm long, inhabits central Asia, northern China and India. It lives beside mountain lakes and streams, often frequenting grass-covered plateaux where it forages for grass, shoots, berries and seeds, and sometimes takes grain from cultivated fields. In winter it descends to the valleys, and leaves the more northerly regions for Burma. At this time it wanders in family groups or in flocks of up to 100 birds. In the breeding season, pairs settle in their nesting grounds, situated up to a height of 4 300 m. The geese build a nest on the ground, among stones in tall grass or on islands scattered over the lakes. The female lays 3—6 whitish eggs, which she surrounds with heaps of down to protect them from the cold. She incubates the clutch for 28—30 days, while her partner stays nearby to guard her. Young goslings are guided by both their parents, the father defending them particularly bravely against predators, and often chasing raptors away.

Argali

JERDON'S BAZA or BROWN-CRESTED LIZARD HAWK
Aviceda jerdoni

Jerdon's Baza, or the Brown-crested Lizard Hawk, is distributed from southern India and the eastern Himalayas across Burma, Thailand and southern Laos to the Philippines, Sumatra, Borneo and Sulawesi. It reaches a length of about 50 cm. The female is paler in colour than the male and has a russet-white chest. Jerdon's Baza inhabits wooded mountain slopes up to a height of 1 800 m. It lives in pairs but forms groups after the fledging of the young. It is active mostly at dusk, flying along the edges of forests in search of lizards. It feeds mainly on lizards of the genus *Calotes*, although it takes tree frogs and larger insects, and sometimes catches small rodents on the ground. Both birds help to build the nest, which is situated 7— 20 m above the ground in a deciduous forest tree. It is made of twigs, and the nesting cup is lined with grass and green leaves. The female lays 2—3 greenish-white eggs, and incubation is carried out by both parents and lasts for about a month.

LAMMERGEIER or BEARDED VULTURE
Gypaetus barbatus

The Lammergeier, or Bearded Vulture, is one of the world's largest raptors. It is over 125 cm long and spans 280 cm. It is a common inhabitant of central Asia and the Himalayas, as well as being found in southern Europe and northern Africa. Outside the nesting season it is usually solitary. It slowly circles the sky, describing wide arcs above the mountain slopes and valleys. It is normally encountered at heights of 1 200—4 000 m, but in spring it goes as high as to 7 500 m. It often cruises above mountain villages and settlements in search of food, for the large carcasses of animals both domestic and wild form the bulk of its diet. The Lammergeier eats small bones and pecks the marrow out of larger ones. It often grasps a large bone such as the femur of a donkey in its beak, soars up to a height of 50—70 m, and drops the bone on a rock, shattering it so that it can feed on the pieces. It allegedly smashes the shells of tortoises in the same way. In the courtship season, the male and female display in acrobatic flights high above the valleys and ravines. They build their nest on a rocky outcrop, constructing it from a heap of branches and scraps of hides, hair and bones. Each pair builds 2—3 nests, and uses them over a period of several seasons. In Asia the breeding season lasts from December to March. The female lays 1 or 2 cream-coloured eggs with russet-brown spots. Incubation is undertaken by both birds for 55—60 days, but

Bar-headed Goose

the female spends more time on the nest. There is usually just a single young, which matures after 5—6 years. Young lammergeiers can be identified by their blackish heads. This bird can make a piercing, whistling call, but it is usually silent.

HIMALAYAN GRIFFON
Gyps himalayensis

The Himalayan Griffon is found in the Himalayas, the Pamirs and in Tibet. This raptor is up to 125 cm long,

*Jerdon's Baza
or Brown-crested Lizard Hawk*

Lammergeier or Bearded Vulture

pack animals can be left along the way, and the griffons immediately swoop down to feed on the carrion. They tear out pieces of flesh, skin and smaller bones from the carcasses before they even have time to get cold. Griffons often cruise above pastures dotted with grazing sheep, for food is always plentiful here. Breeding takes place between January and March, the birds building a nest on an inaccessible rock wall, using branches broken off with their beaks. The nesting cup is lined with scraps of hide. A colony of 5—6 pairs sometimes nests close together. The female lays a single egg, usually white, but occasionally with russet specks. Both partners incubate it for about 50 days, and the young griffon stays on the nest for 80 days, being fed on food regurgitated from its parents' crops.

BLACK VULTURE
Aegypius monachus

The Black Vulture is a resident of mountainous areas from Asia Minor to central Asia and Tibet. It also occurs in southern Europe. It reaches a length of 110 cm and has a wingspan of 265—287 cm. It seeks open country, usually at heights between 2 400 and 3 000 m. In mid-February vultures build large nests of branches in the trees, pairs sometimes using the same nest for many seasons, and enlarging it every year. Some nests reach a height of 2 m and a diameter of 1,5 m. In Asia, the Black Vulture usually builds its nest in a tall juniper tree, 8—12 m above the ground. Vultures use their powerful beaks to break off branches for nesting material. The female lays a single, pale yellow egg with reddish-brown spots, which is incubated by both birds for 55 days. The newly hatched chick is fed on food regurgitated from its parents' crops. It stays in the nest for 90—105 days, by which time it is capable of taking its first flight. The Black Vulture feeds mainly on large carrion, but it occasionally hunts live prey, such as reptiles, amphibians and small mammals.

and has a wingspan of 250 cm. It is a resident of open mountain valleys and rocky slopes at heights from 600—2 500 m, but it sometimes seeks its food even at altitudes in excess of 4 500 m. Griffons live usually singly or in pairs, though they may form small groups. The birds circle continuously above the rocks and ravines, sometimes following caravan routes for hours. Their patience often brings its reward, for dead

Himalayan Griffon

IMPERIAL EAGLE
Aquila heliaca

The Imperial Eagle is widespread from southern Europe to Lake Baikal. In Asia it is common in northern India and China. Outside the nesting season, it migrates to Vietnam and Laos. It reaches a length of 90 cm and a wingspan of 180 cm. This eagle inhabits forests on the lower slopes of mountains, low-situated mountainous areas and forests, where it builds nests of twigs and sticks in the trees, 6—9 m above the ground. Pairs often use the same nest for many seasons, repairing and enlarging it every year. They usually have 2—3 nests in the same neighbourhood and use them intermittently. The female lays 2 white eggs with reddish-brown spots. Both birds incubate the clutch for 42—45 days and feed the young for 65—67 days on the nest. They continue to bring food for some time after the young have fledged. The Imperial Eagle mainly hunts small mammals, but it takes birds and reptiles as well, including venomous snakes. In India, it catches extremely poisonous species of viper. It seeks its prey from a high vantage point or while cruising above the countryside, and it often eats carrion.

BLYTH'S HAWK-EAGLE
Spizaetus alboniger

Blyth's Hawk-eagle, which is about 70 cm long, inhabits the open hilly woodland of Thailand, Malaysia, Sumatra and Borneo, usually at heights from 400—1 800 m. During the nuptial period it emits whistling metallic sounds. The nest is built in huge trees such as the Sal Tree *(Shorea robusta)* 12—25 m above the ground, and consists of a great number of branches and a lining of green leaves. Each pair normally has two nests, using them in alternate seasons and repairing them regularly. The female lays 1 or sometimes 2 white eggs, occasionally patterned with a few yellowish or reddish dots. The female incubates the clutch for 40 days while the male brings food to her. After hatching has taken place, he feeds the whole family, bringing food to his mate for her to share out. Blyth's Hawk-eagle preys on fowl-like birds, particularly pheasants. It also visits the edges of villages, where it takes chickens.

LESSER FISHING EAGLE
Ichthyophaga nana

The Lesser Fishing Eagle is native to the Himalayas, south-eastern Asia, Sumatra, Borneo and Sulawesi. This raptor is about 65 cm long. It inhabits wooded

Black Vulture

Imperial Eagle

Blyth's Hawk-eagle

Lesser Fishing Eagle

locations beside mountain streams at a height from 1 000—2 400 m. It perches on branches overhanging the water and waits for fishes, which form the bulk of its diet. It seizes its prey in the water, with its long claws. Birds or small mammals on the banks are occasionally caught as well. In the breeding season, which takes place between March and May, it makes loud, jingling cries. The Lesser Fishing Eagle makes its nest of branches and sticks in the crown of a tall tree. It is used for many years and is gradually enlarged. The nesting cup is lined with a large quantity of green leaves. The clutch consists of 2—3 white eggs.

CHUKAR PARTRIDGE
Alectoris chukar

The Chukar Partridge is distributed from Bulgaria to central Asia, from where it extends to Mongolia. It frequents rocky locations in the mountains, up to a height of 4 000 m. It is a common bird, much hunted locally. The Chukar Partridge runs speedily and nimbly on the steep slopes. When disturbed, it scurries uphill and then glides back down into the valley, so easily escaping its enemies. It makes a sharp warbling call as it runs, to warn other birds of danger. For most of the year, this partridge lives in family groups, or gathers in large flocks in autumn. It forms pairs in the breeding season. The female builds a nest in a depression on the ground, lining it with pieces of plant material. The nest is hidden away under a low bush or projecting rock, often in a tuft of grass. The female lays 8—15 eggs, yellowish-beige in colour, with tiny brown dots. She incubates the clutch for 25—26 days, and both parents care for the chicks. This species feeds on insects, insect larvae and other small invertebrates, and on seeds, berries and green plants. The Chukar Partridge is preyed upon by raptors and small carnivores which often take chicks which are not yet able to fly.

SNOW PARTRIDGE
Lerwa lerwa

The Snow Partridge is distributed from Afghanistan across the Himalayas to southern Tibet and western China. It reaches a length of 40 cm and both sexes are alike in colour. It is a resident of mountainous areas at heights of 3 000—5 000 m, occasionally descending in winter to an altitude of 2 500 m. The Snow Partridge seeks localities where there is a thin cover of grass, low rhododendron bushes and permanent snow patches. It lives in families, usually of 6—8 birds, forming coveys of up to 25 in winter. It forms pairs in the breeding season. In May or June, the female builds a nest in a depression under an overhanging rock or beneath a bush on a steep slope. She lines the nest with moss and leaves and lays 3—6 beige eggs with reddish spots. The duration of incubation is unknown, but both parents have been seen to rear the chicks. Snow partridges feed on lichen, moss, seeds and shoots, and in spring and summer they also feed on insects, insect larvae and other small invertebrates. In the nesting season the Snow Partridge makes a melodious giggling call. On sensing danger, it produces a sharp whistle.

COMMON HILL PARTRIDGE
Arborophila torqueola

The Common Hill Partridge is found in the Himalayan region from which it extends to Burma, Malaysia and northern Vietnam. It reaches a length of about 30 cm. The female resembles the male, but she has a rust-coloured chin and throat, dotted with black, and a brown breast. This partridge lives in thinly wooded areas at a height of 1 500—4 000 m. During the day, it combs the soil in search of larvae, worms, molluscs and spiders. It also feeds on berries and shoots. At night, it roosts in the trees. For most of the year partridges move in small groups of 5—10, but they form pairs in the breeding season. The female makes a shallow scrape under a low bush or at the foot of a tree, lining it with grass. She lays 3—5, or even up to 9 eggs, usually whitish in colour. She incubates the clutch alone for about 24 days.

Chukar Partridge

Snow Partridge

Common Hill Partridge

Caspian Snowcock

CASPIAN SNOWCOCK
Tetraogallus tibetanus

The Caspian Snowcock is native to Tibet, where it inhabits mountain slopes at heights from 3 500—5 500 m. It is up to 70 cm long. The female resembles the male in appearance, but she has yellowish spots on the greyish-white band across her breast. For most of the year the Caspian Snowcock travels in small

Himalayan Monal Pheasant

flocks, forming pairs only in the nesting season, at which time the male often makes loud, whistling sounds, usually of five notes. The female builds a nest on the ground, among stones on a steep slope or in a clump of thin grass, and lines the depression sparsely with grass stalks or leaves from bushes. She lays a clutch of 6—9 spotted eggs and incubates them for about 4 weeks. The chicks are reared by both parents and stay in their company until September, when the families merge to form larger flocks for the winter. Snowcocks eat green plants, moss, roots and a variety of berries. Only the chicks feed on insects and insect larvae. When pursued, snowcocks run uphill, then turn back and glide across the valley to the opposite hillside. The high Asiatic mountains are the home of various other species of snowcock.

HIMALAYAN MONAL PHEASANT
Lophophorus impejanus

The Himalayan Monal Pheasant covers a range from eastern Afghanistan to Bhutan and southern Tibet. The male reaches a length of 70 cm and a weight of 2—2.4 kg. The female is smaller, weighing about 2 kg and being predominantly greyish-brown in colour, but fawn below, and with a white band on the throat. The Himalayan Monal Pheasant is a common game bird of the mountains at heights from 2 600—5 000 m. It usually moves in small groups of 3—4 made up of a cock with 2—3 hens, but in winter, pheasants form coveys composed of either hens or cocks. The nesting season lasts from April to June. The female makes a scrape under a bush or fallen trunk, lining it with leaves, pine needles, grass or feathers. She lays 4—6 buff or russet-coloured eggs, densely covered with fine reddish dots, and incubates the clutch for 29—30 days. In their first year of life, young cocks resemble the hen, though they sometimes have a few black feathers, only becoming fully coloured during their second year. The Himalayan Monal Pheasant feeds on seeds, berries, shoots, roots, grass, insects, worms and other invertebrates. The birds often dig out their prey with their curved beaks, and in winter, they frequently shovel the snow to reach the roots underneath, spending hours turning the soil over. These birds seek the cover of birches or rhododendrons, or thin oak woodland, descending to the valleys in winter.

SATYR TRAGOPAN or CRIMSON HORNED PHEASANT
Tragopan satyra

The Satyr Tragopan, or Crimson Horned Pheasant, is indigenous to the Himalayas and southern Tibet. The male reaches a length of 70 cm, while the female is about 60 cm long and is covered with both dark and light brown spots. Young cocks assume the adult coloration in their second year. Adult males weigh up to 2.1 kg, and females reach a weight of 1.2 kg. This species inhabits forested slopes at a height of 2 400—4 250 m, descending to about 1 800 m in winter. Courtship displays begin in March, the male spreading his colourful breast feathers and raising the blue horns above his eyes, to impress the females. His mate builds her nest on the ground, or in nests among the trees vacated by other birds. She lays 2—4 eggs in a tree nest, or 6—7 eggs on the ground. They are a reddish-yellow colour, with fine russet dots, and are incubated by the female for 25—28 days. The parents feed their young on green leaves and insects. The chicks take the insects directly from the adults' beaks. This applies particularly to the young hatched in the trees. These are well-developed and embark on their first flights after only a few days, roosting in the trees with their parents. Satyr tragopans feed on the leaves of various plants, and even young birds are capable of swallowing extremely large leaves. They also eat shoots, berries, roots, seeds, insects and other invertebrates.

BLOOD PHEASANT
Ithaginis cruentus

The Blood Pheasant is distributed from the Himalayas and Tibet to Kansu, Yunnan and north-eastern Burma. The male is about 47 cm long, while the less colourful female measures 40 cm. The male has two or more characteristic spurs on each tarsus. This species dwells in the high mountains, living at heights of 3 600—4 500 m, and sometimes ascending to the snowline. In winter it settles in thin pine woods, moving to dense rhododendron cover in summer. Flocks of 5—10, or sometimes up to 30 birds, roam the

Satyr Tragopan or Crimson Horned Pheasant

Blood Pheasant

countryside in winter. Blood pheasants run swiftly among the rocks on the grassy slopes. They rarely fly, except when danger threatens. Then they take off and fly down a slope, although they always walk uphill. In the nesting season they normally form pairs, although in some subspecies, males have been seen to live with two hens, and on occasion a hen has been observed in company with two cocks. In April or May the female builds a shallow depression under a rhododendron shrub, lining it with grass stalks and leaves. She lays 5—12 eggs, pinkish with brown spots, and incubates them for 28—29 days. The Blood Pheasant feeds on

Blue-eared Pheasant

berries, lichens, mosses and shoots growing on stunt-ed birch and pine trees. Young birds also eat insects and insect larvae. Blood pheasants often forage in open situations but they always stay near dense undergrowth where they can take shelter. In some areas, they are commonly sought as game.

BLUE-EARED PHEASANT
Crossoptilon auritum

The Blue-eared Pheasant obtains its name from its elongated head feathers. Interestingly, the sexes are alike in colour in all pheasants of this genus and the vanes on their tail feathers look frayed. The short tail feathers contributed to a drastic reduction in number of these birds, for they were used for ornamentation and as symbols of dignity on the hats of Chinese mandarins. The Blue-eared Pheasant inhabits the mountainous areas of the Chinese provinces of Kansu, Tsinghai, and the mountains around Ala Shan. It is a typical bird of the high mountains, living in wood-

land and on grassy slopes at heights from 1 500—4 000 m. It favours localities with streams, where it drinks regularly. In the breeding season, the Blue-eared Pheasant lives in pairs, and the cocks make characteristic piercing calls, mainly in the morning and evening. They stand on elevated sites and their voices carry far afield. The courting displays of blue-eared pheasants differ from those of other kinds of pheasants. While most male pheasants whir their wings, the cock of the Blue-eared Pheasant struts around the hen with his nearer wing hanging low and dragging half of his tail. The hen builds a nest in dense undergrowth while the male stays nearby to warn her of danger. The female lays 5—12 light brown eggs in a depression lined with grass and leaves. She incubates the clutch alone for 26—28 days, occasionally leaving for half an hour to eat and drink, but always carefully covering the eggs with grass and leaves when she does so. When the brood is fledged the families form flocks of about 30 and roost together in the trees. In the morning they fly down to forage on the ground. They use their stout beaks for digging, and pecking at worms, molluscs, insect larvae and the roots of grasses and various other plants. They also take small vertebrates, killing them with their beaks, and they collect seeds or pluck shoots and berries from bushes and trees. Unlike other fowl-like birds, they never use their feet for digging. Blue-eared pheasants reside always in the same area. They usually settle on a particular hill, and when flushed they race uphill and then glide down again from the top. Their flight is very clumsy and they prefer to run. As well as being hunted for game they are preyed upon by leopards, foxes and raptors, especially hawks.

WHITE-EARED PHEASANT
Crossoptilon crossoptilon

The White-eared Pheasant ranges from eastern Tibet to north-western Yunnan and in the south to north-eastern Assam. There are five subspecies, differing in coloration but they are all about 95 cm long, and have velvety black crowns, naked red cheeks and white ear feathers. This species lives in the high mountains at heights from 3 000—4 500 m, or even higher. In winter it descends to woods at lower altitudes, though almost never below 3 000 m. In the nesting season it forms pairs living in dense rhododendron groves or in thin woodland. It forages in open localities in the

White-eared Pheasant

Silver Pheasant

morning or late afternoon, though always near protective undergrowth. The female lays 4—9 eggs in a shallow depression lined with greenery. They are pale grey or brownish in colour, sometimes with a few russet spots. She incubates the clutch for 24—26 days. After fledging, the young pheasants live with their parents in families which merge to form small flocks in autumn. Their diet is similar to that of the Blue-eared Pheasant.

CHEER PHEASANT
Catreus wallichii

The Cheer Pheasant is native to western and central areas of the Himalayas where it lives at a height of 1 300—3 300 m. The male is up to 110 cm long, and the female measures about 75 cm. She is similar to the male in coloration except that her underparts are chestnut brown with small spots. For most of the year these pheasants live in small groups of 5—7, probably made up of members of the same family. They roam over the steep slopes in search of food, mainly roots, tubers, seeds, berries, insects, worms and molluscs. They often dig in the ground with their powerful beaks, pecking out both animal and plant food, and in spring and summer they pluck grass and green shoots. In the nesting season the pheasants form pairs, and the female builds a nest under a thick bush and lines it with leaves and grass. She lays 7—14 creamy white to pale brown eggs with reddish-brown spots, and incubates them for 26 days.

SILVER PHEASANT
Lophura nycthemera

The Silver Pheasant is widespread from the Himalayan region to south-eastern Asia. There are 13 subspecies, differing in intensity of coloration and in the width of the bands on the tail feathers. The male is up to 125 cm long including the tail, which measures 75 cm. The female is 60 cm long, and is predominantly brown above and greyish-brown below with dark-rimmed feathers. The Silver Pheasant lives in mountain areas at heights from 700—2000 m, its favourite haunts being among thin woodland with dense undergrowth. The male has a harem of 2—4 hens. The female lays 10—18 pinkish eggs, sometimes with white

Cheer Pheasant

Lady Amherst's Pheasant

spots, in a nest either on the ground under a bush, or in the grass beneath a tree. She incubates the clutch by herself for 25 days, and then rears the chicks. This pheasant feeds on seeds, berries, green plants, fruit, insects and other small invertebrates, and even on small vertebrates such as geckos and young snakes.

LADY AMHERST'S PHEASANT
Chrysolophus amherstiae

Lady Amherst's Pheasant is distributed from western China to north-eastern Burma, being found at heights from 2 000—5 000 m in the mountains. The male reaches a length of up to 170 cm, but 110 cm of this is taken up by the elongated tail feathers. The female is greyish-brown with narrow, black, wavy lines, and reaches a length of 90 cm. The cock has a harem of 3—4 hens. The female builds a nest in a depression under a bush, lining it with grass and leaves. There she lays 6—12 cream-coloured eggs and sits on them for about 23 days. She rears the chicks and watches over them by herself. This species feeds on seeds, green plants, berries and various small invertebrates,

mostly insects and insect larvae. The young of small vertebrates, such as lizards or rodents, are occasionally also taken.

IBISBILL
Ibidorhyncha struthersii

The Ibisbill inhabits rocky banks of mountain rivulets and streams from Turkestan across the Himalayas to northern China and Mongolia. This remarkable bird lives at heights from 1 700—4 400 m, descending to the valleys in winter and migrating from the more northerly regions in small flocks of 6—8 to spend the winter in the mountains of north-eastern Burma. It seeks its food under stones in shallow water pushing its long bill so far down that the bird is often submerged up to its breast. It also catches aquatic insects and their larvae, crustaceans and molluscs. In the courtship season, it produces a whistling, repeated flight call. Pairs form and defend their permanent territories. They build a nest in a shallow depression in thick grass near a bank, lining it with dry stalks. The clutch usually averages 3 spotted eggs. Incubation is shared by both birds, but the female spends more time on the nest than the male. The young hatch after 28 days and leave the nest as soon as they are dry, to follow their parents. They begin to fly when they are about 40 days old.

ROCK DOVE
Columba livia

The Rock Dove is native to the whole of south and south-east Asia and it is also found in southern Europe. In addition, it has been successfully introduced and has become acclimatized in many places. Rock doves, which are about 35 cm long, abound in the

Ibisbill

cities of south-eastern Asia, living on tall buildings and often cross-breeding with domestic pigeons. Their favourite haunts are rocky sites, particularly along valleys, where they live in colonies of over 100 birds. In Nepal and Bhutan, they occur at heights up to 3 000 m. Pairs build their nests in small caves and rocky crevices, assembling them from a few sticks and grass stalks. In towns rock doves settle on beams in attics, or in cracked walls, and a pair may occupy a hole in a palm tree. The female lays 2 white eggs and the partners take turns in incubation for 16—17 days. At first they feed their young on moist food regurgitated from the crop, but later they feed them on pre-digested seeds. Young doves leave the nest after 3 weeks, but the parents continue to feed them for some time. Adult birds forage in the morning and evening, mainly for seeds and green plants, while flocks of rock doves frequently raid cornfields or rice paddies and cause considerable damage to the crops. Their principal enemies are raptors, especially hawks. The Rock Dove is the ancestor of the domestic pigeon.

HIMALAYAN SNOW PIGEON
Columba leuconota

The Himalayan Snow Pigeon, which is about 35 cm long, lives in the high mountains of the Himalayan region, at heights from 3 000—5 000 m. It occurs in colonies which settle on steep rock walls, and flocks of more than 150 pigeons forage for food on grassy slopes, often on the edges of the snowfields. They feed on grass seeds, grain, various berries and green plants, especially polygonums. In summer the pigeons often raid cultivated land near human dwellings. In the breeding season these birds make crowing sounds, quite different from the calls of other species of pigeon. Each pair builds a flat nest made of small sticks and feathers, on a sheltered rocky ledge or in a small cave in a rock wall. The clutch of 2 white eggs is incubated by both partners for 17 days, and the young are reared by both their parents.

IMPERIAL PIGEON
Ducula badia

The Imperial Pigeon has an area of distribution covering wooded mountainous regions from India and the Himalayas to south-western China, and across Thai-

Rock Dove

Himalayan Snow Pigeon

land to the Greater Sunda Islands. This pigeon measures up to 50 cm. It keeps to the trees on slopes up to a height of 2 400 m. The bulk of its diet is composed of various fruits, especially wild figs, and the fruits of nutmeg trees, which the birds swallow whole. Small flocks of 15—20 sometimes fly long distances to places where berries and fruit are abundant. In the early morning and afternoon the pigeons bask in the sun in the treetops, and they have regular sites in massive trees where they roost at night. Nesting takes place between March and August, depending on locality. The nest is built among branches, 5—8 m above the ground, and is made of twigs. The female usually lays a single white egg, which the parents incubate for 17 days. Imperial pigeons often nest in colonies, the nests being only about 1 m apart. When the

Imperial Pigeon

south-eastern Asia, Sumatra and Java. It is about 40 cm long. The female has dark brown striping on her head and breast. This species lives in pairs which merge to form small groups outside the nesting season. It seeks its food in the treetops, mainly collecting small fruits and berries, and pecks seeds on the ground. It is silent for most of the year, but makes a deep lowing call in the courtship season. Each pair builds a flat nest of twigs in a deciduous tree, 2—8 m above the ground. The female usually lays a single egg and sits on it alternately with the male for 17 days. The chick is fed by both parents.

broods are fledged, groups of pigeons frequently undertake long, wandering trips in quest of food, returning to breed in their original habitats.

BARRED CUCKOO DOVE
Macropygia unchall

The Barred Cuckoo Dove inhabits mountain forests at heights from 450—2 700 m, ranging from Kashmir to

TIBETAN SANDGROUSE
Syrrhaptes tibetanus

The Tibetan Sandgrouse, which is about 45 cm long, inhabits the Tibetan plateaux, central Asia and the Kirgiziya Steppe. The female differs from the male in having black wavy lines on her back. In Tibet this sandgrouse occurs at heights of 4 200—5 400 m. It lives on the ground in places where there is almost no water, and it has to fly many kilometres in order to drink at water-holes, springs and streams early in the morning and before sunset. It is sociable, travelling in flocks of 10—30 or more outside the nesting season. It pecks up seeds in grassy localities, plucks green shoots and occasionally catches beetles. In the breeding season, the pairs defend small territories and build nests near to one another. The nest consists of a shallow depression sparsely lined with dry stalks. Nests on rocky sites are not lined. The female lays 3 pale yellow eggs, with brown and violet spots. She sits on them during the day, being relieved by the male at night. The cock often brings water for the hen in his crop. The young hatch after 28 days and both parents feed them and bring them water.

RED-BREASTED PYGMY PARROT
Micropsitta bruijnii

The Red-breasted Pygmy Parrot inhabits the Moluccan islands of Buru and Seram, New Guinea, the Bismarck Archipelago and the Solomon Islands. It reaches a length of 9 cm. The female is green, with a blue crown and her cheeks are orange and white. This species is abundant in mountainous areas at heights from 500—2 300 m, living in pairs which

Barred Cuckoo Dove

merge to form small groups when the young are fledged. Pygmy parrots climb over branches and tree trunks, gripping with their long claws as they search for lichens, fungi and insect larvae in the crevices of the bark. Like woodpeckers they carve out a hole in the rotting wood of a stump or dead branch about 3 m above the ground. The hole is about 4 cm deep and the birds sleep in it. Here also the female lays 3—4 eggs and sits on them for 18 days. The young are fed by both parents.

HIMALAYAN SLATY-HEADED PARAKEET
Psittacula himalayana

The Himalayan Slaty-headed Parakeet is distributed in eastern Afghanistan and Kashmir, from where it extends across the Himalayas to Assam. It is about 40 cm long. The female differs from the male in that her head is a lighter colour and she lacks the red spot on the upper mantle. This parakeet favours woodland on mountain slopes at heights from 250—3 800 m. Outside the nesting season, it moves in small groups, foraging for seeds and various fruits and shoots. The nest is built in a hole in a tree. The female lays 4—5 eggs and incubates them for 22 days, while the male feeds her. The young are fed by both parents for 6—7 weeks on the nest and for some time after they are fledged. Two or three parakeet nests are sometimes found in the same tree. When the brood is fledged, the family roams throughout the surrounding area and later joins up with other family groups.

GOLDEN-MANTLED RACKET-TAILED PARROT
Prioniturus platurus

The Golden-mantled Racket-tailed Parrot lives on Sulawesi and the adjacent islands and on Talaud Island. It attains a length of 28 cm. The female is distin-

Golden-mantled Racket-tailed Parrot

Rajah's Scops Owl

guished from the male by shorter tail feathers and a greenish abdomen. This parrot is a resident of mountain forests at heights from 1 800—3 000 m. It seeks food in the treetops, eating ripe fruit, berries and seeds and pecking green shoots. The nest is built in a tree hollow, either in the trunk or in a large dead bough. Little is known about the breeding behaviour of this parrot.

RAJAH'S SCOPS OWL
Otus brookei

Rajah's Scops Owl, which is about 25 cm long, is found in southern, south-eastern and eastern Asia, and on the Greater Sunda Islands and the Philippines. It dwells in pine or oak woods at heights from 900—2 400 m in the mountains, and its monotonous hooting call can often be heard in city parks and gardens. During the day it hides in dense treetops or in hollow trees. It comes out to hunt after sunset, catching insects and small birds. It builds its nest in holes in trees, usually in palm trees, and it occasionally settles in crevices in the walls of buildings. The female seldom lines the nest, but lays 2—3 eggs directly in the bottom of the hole. Then she sits on the clutch for 26 days, while the male brings food to her. These owls bravely defend their young against predators, often attacking an intruder with their sharp claws, and chasing away animals more powerful than themselves. They have even been known to charge at men approaching their nest.

BROWN WOOD OWL
Strix leptogrammica

The Brown Wood Owl is widely distributed over south and south-east Asia and on the Greater Sunda Islands. This common owl, which is about 50 cm long, frequents dense mountain forests, usually at heights from 750—2 500 m, although in the Himalayas and other high ranges it can be seen at 4 000 m. It sleeps during the day, hidden among dense branches in the treetops, or in a hole in a tree. It comes out to hunt at twilight, flying expertly through the canopy and catching birds up to the size of domestic fowl. It also preys on small mammals and lizards. Nesting takes place in the first half of the year, usually between January and March. The female lays 1—2 white eggs in a hole in a tree or in a rocky crevice, and incubates the clutch for 28—30 days, while the male feeds her. When the young are hatched, the female stays with them all the time for the first few days, and the male looks after the whole family. When they are 35 days old, the young owls leave the nest and perch on nearby branches, where their parents bring prey to them. After 50 days, they take their first flights and learn to hunt.

BLACK COUCAL
Centropus toulou

The Black Coucal ranges from India across southern China to Taiwan and Hainan, and to the Philippines, the Greater Sunda Islands and the Moluccas. This species is characterized by its long tail. The male is about 35 cm long, and the slightly larger female measures about 40 cm. The Black Coucal is resident in most areas, but the most northerly populations migrate to Thailand to spend the winter. Its favourite haunts are in the foothills of high mountains, although in the Himalayas it can be seen at a height of 2 000 m. This coucal frequents localities covered by bushes and tall grass. It moves swiftly among the dense vegetation, hunting for insects, spiders, molluscs and other invertebrates, and occasionally catching small lizards. For most of the year it lives in pairs which defend their territories and build their own nests. The nest is a spherical construction with a side entrance, made from roots, grass stalks, twigs, leaves and slivers of wood. It is lined with small green leaves. It is usually situated 1—1.5 m above the ground in a bush, but it is sometimes built in the crown of a palm tree. The female lays 2—4 pure white eggs and incubates them for about 18 days. It is not

Brown Wood Owl

known whether the male assists her. The flight of the Black Coucal is slow and clumsy, and it never flies over open locations from choice, since it easily falls prey to raptors in the open.

MIGRATORY or JUNGLE NIGHTJAR
Caprimulgus indicus

The Migratory or Jungle Nightjar ranges from south and south-east Asia to the Philippines, the Greater

Black Coucal

Migratory or Jungle Nightjar

Ceylon Frogmouth

and other insects, which it swallows whole. In the courtship season, it makes deep growling sounds, repeated as many as fifty times. The female lays 2 pinkish-white eggs with a few brown spots on the bare ground, and incubation is carried out by both partners for 17 days. The chicks are fed on insects brought by both parents for the first 18 days. After this time they begin to fly, but the adults still feed them for a few days longer. The nesting period varies according to the area of distribution. In Sri Lanka it takes place between February and June.

CEYLON FROGMOUTH
Batrachostomus moniliger

The Ceylon Frogmouth, which is about 25 cm long, is found in the mountains of Sri Lanka, at heights from 300—1 800 m. It is a bird of the dense primary forests, and on account of its strictly nocturnal life, it has been seen by very few people. During the day, it perches across a branch, never lengthwise like the true nightjars to which it is related. When it needs to look behind, it turns its head right round in the same ways as owls. The frogmouth sleeps deeply and can be easily caught in the hand before it wakes up. After nightfall it flies among the trees, catching insects in its open beak, especially beetles which it swallows whole. Nesting takes place between February and March, and again in September. The nest is situated on a horizontal branch and is made from bark, moss and lichen. The construction measures about 6 cm across and 1.5 cm deep, and is lined with the birds' own down. The female lays a single white egg, which is incubated by the male during the day and by the female at night.

Sunda Islands and New Guinea. It inhabits sparsely wooded regions and bush-covered mountain slopes up to a height of 3 300 m. It is solitary, forming pairs only in the breeding season. By day it rests on a branch or in the grass under a bush, and when it is flushed, it flies up quickly and quietly, landing again a few metres further on. At twilight it flutters just above the grass in search of butterflies, beetles, locusts

LARGE BROWN-THROATED SPINETAIL SWIFT
Hirundapus giganteus

The Large Brown-throated Spinetail Swift is native to south and south-east Asia and to the Greater Sunda Islands, but it does not nest in Sri Lanka, as it is never there during the breeding season. This bird reaches a length of 25 cm, and dwells in forests at heights of up to 2 000 m. It usually lives in large flocks, flying at great speed above woodland, rivers and tea plantations. While in flight, these birds often spread the thorn-like tail feathers from which they get their

Large Brown-throated Spinetail Swift

name. Swifts catch insects in the air as they fly. Each pair builds a nest in a tree hollow vacated by woodpeckers, or in a rocky cleft. The female lays 3—5 white eggs at the bottom of the hole and sits on them for 19 days. The young are fed by both parents.

EDIBLE-NEST SWIFTLET
Collocalia fuciphaga

The Edible-nest Swiftlet inhabits southern areas of south-east Asia, the Sunda Islands, Palawan, the southern Philippines, Sri Lanka and southern India. This species is renowned for its nests, which are used to make the famous 'bird's nest soup'. It occurs in several subspecies. It lives in areas rich in rocks and caves, ranging from low-lying land, to an altitude of 3 600 m. The Edible-nest Swiftlet lives in large colonies and in some places it is the most common of the local birds. It flies very quickly, catching flies and swarming ants on the wing. The colonies nest in total darkness in deep caves or crevices among the rocks. The swiftlets are able to fly without hitting the rock walls by means of echolocation, for they make short sharp rattling sounds which echo from the walls. Thousands of nests are often heaped close to each other or stuck in layers to the walls and roof. The swiftlets use saliva as the main building material, sometimes adding moss, lichen or other delicate plant material collected from branches and hanging lianas in the forest. The nest is thick at the base and is shaped like a shallow cup. The saliva rapidly hardens to the consistency of gelatine, and the nest can be eaten after being boiled in water. The female lays 2 white eggs, which are incubated by both partners for about 16 days. The young are fed by both parents.

RED-HEADED TROGON
Harpactes erythrocephalus

The Red-headed Trogon is distributed from Nepal to south-eastern Asia and to Sumatra, inhabiting mountain forests up to a height of 700—1 800 m. It is about 35 cm long. The female has a brownish-orange head, neck and breast. The Red-headed Trogon is solitary, forming pairs only in the nesting season. It preys on airborne insects. In the courtship period the male makes whistling sounds, repeated 5—6 times. The nest is built by both parents in a hole in a tree.

Edible-nest Swiftlet

Red-headed Trogon

Greater Pied Kingfisher

The female lays 2—5 white eggs and incubates them for about 18 days alternately with the male. The young are fed by both their parents.

from Kashmir across China to Japan and south-eastern Asia. It attains a length of about 40 cm. The female resembles the male, but has rust-coloured feathers beneath her wings, which show when she is flying. This kingfisher frequents rivulets and streams in mountainous areas up to a height of 2 000 m. It catches fishes in the crystal-clear water, but it also preys on insects and other aquatic invertebrates. It often flutters in one spot over the water, scanning the surface. When it pinpoints its target, it swoops down like a stone, with its wings folded. The Pied Kingfisher usually lives in pairs. In the nesting season each pair digs out a burrow over 1 m long in a clay or sandy bank, terminated by a nesting chamber. The female lays 2—5 white eggs directly on to the bottom of the chamber, and incubates them for about 22 days. The young hatch out one at a time, for the female sits on the nest from the time the first egg is laid. She is occasionally relieved by her mate. The chicks are born blind and almost naked, and their parents feed them on insects and tiny fishes. Young kingfishers leave the nest when they are 25 days old, but the adults feed them for another 10 days, while they teach them to hunt.

GREATER PIED KINGFISHER
Ceryle lugubris

The Greater Pied Kingfisher is widely distributed

Ruddy Kingfisher

RUDDY KINGFISHER
Halcyon coromanda

The Ruddy Kingfisher, which reaches a length of about 25 cm, is found in southern and south-eastern Asia, and in eastern China and Japan. It lives near brooks, rivulets and lakes at heights up to 2 000 m in the forests. It is very cautious and always perches among dense foliage, where it becomes virtually invisible, though its laughing voice is often heard. This kingfisher lives singly for most of the year, and forms pairs only in the breeding season. The nest is built in a tree hollow abandoned by woodpeckers. The clutch consists of 2—3 white eggs, but very little is known about the nesting habits of this beautiful species. The Ruddy Kingfisher feeds on small fishes and insects.

GREAT SLATY WOODPECKER
Mulleripicus pulverulentus

The Great Slaty Woodpecker ranges from northern India to south-western China and across south-eastern Asia to the Greater Sunda Islands and Palawan. It

is about 50 cm long, and the female lacks the red
stripe stretching from the beak to behind the eye. This
woodpecker is a resident of evergreen mountain forests, usually at a height of about 1 000 m, but it can
also be seen as high as 2 000 m. Its flight is more like
that of crows than of other species of woodpeckers. It
often makes loud cries when in flight. Like other related species, it hammers branches of trees with its beak
during the nuptial season. It usually lives in pairs,
which chisel out burrows, up to 60 cm deep, in large
tree trunks. This task takes about 10 days even if the
tree is dead. The same hole is often used for several
years. The female lays 2—4 eggs and sits on them alternately with the male for about 14 days. The young
remain in the nest for 24 days, and their parents feed
them on ants and the larvae of wood-boring beetles.
Adult woodpeckers themselves feed mainly on these
beetles and their larvae. After the brood is fledged,
the family often wanders as a group for a few weeks.

GOLDEN-THROATED BARBET
Megalaima franklinii

The Golden-throated Barbet dwells in mountain forests from Nepal to southern China and south-eastern
Asia. It measures about 23 cm. This bird lives always
on mountain slopes covered with moist primary forest
at heights from 600—2 400 m. During the day it
searches for food in the tall treetops, feeding on soft
fruit, especially figs, and catching insects and insect
larvae. At midday it rests on a branch, and makes its
repeated monotonous call. Barbets nest in small holes
chiselled in rotten wood, often in a dead tree, or in
nests abandoned by woodpeckers. The clutch of 2—3
white eggs is incubated by both partners for 17 days,
and the young are fed by both their parents.

BLUE-NAPED PITTA
Pitta nipalensis

The Blue-naped Pitta has an area of distribution
which covers Nepal, south-western China, Burma and
northern regions of Laos and Vietnam. It inhabits
wooded mountain slopes up to a height of 2 000 m. It
attains a length of 25 cm, and the female differs from
the male in having a golden-yellow nape. Pittas live
singly and form pairs before mating. They are found
among bushes or in thick bamboo jungles, moving

Great Slaty Woodpecker

Golden-throated Barbet

Blue-naped Pitta

briskly in long leaps among the vegetation and fallen leaves where they forage for ants, beetles and other small invertebrates. In the nuptial period the male sings with a melodious voice, mostly in the morning and before sunset. A spherical nest with a side entrance is built by both birds, from leaves, stalks and roots. It is usually situated in the branches of a tall bush. The female lays 3—5 spotted eggs, which are incubated by both parents for 14 days. Both adults rear their brood.

RUFOUS-BELLIED NILTAVA
Niltava sundara

The Rufous-bellied Niltava is distributed from the Himalayas to south-eastern Asia. It reaches a length of 15 cm. The female is brownish-olive above with a blue spot on each side of the neck, a rufous tail and a white spot on the breast, her underparts being fawn in colour. This species frequents bush-covered mountain slopes at heights from 1 800—3 200 m, but descending to below 1 000 m in winter. It usually lives in small flocks during the winter, but some niltavas remain solitary, and only form pairs in the breeding season. The nest is made of moss, and is built among dense foliage in shrubs. The female lays 3—5 eggs with rust brown spots and sits on them for 13 days. The young are fed by both parents. Niltavas feed mainly on insects, skilfully captured on the wing, although insect larvae and other small invertebrates are also collected on branches.

WHITE-CAPPED REDSTART
Chaimarrornis leucocephalus

The White-capped Redstart inhabits rocky and mountainous areas from Turkestan to China, north-eastern Burma, northern Laos and northern Vietnam. It reaches a length of 19 cm and both sexes are alike in colour. It lives at heights from 1 800—5 100 m, descending in winter to the shelter of the valleys. It is solitary for most of the year, forming pairs only in the nesting season. Its favourite haunts are along the rocky banks of mountain rivulets and streams. Here it can scan the water surface from the top of high boulders and dart down to catch insects trapped in the water. It also catches airborne beetles, flies and butterflies, or collects small molluscs, and it also pecks up

Rufous-bellied Niltava

White-capped Redstart

soft berries. In the nesting season, the male produces a long, melodious whistling call. Redstarts look for crevices in rocks in which to build their nests. They are made from twigs, grass stalks, moss and lichen, and are lined with animal hair. The female lays 3—5 eggs and incubates them for about 14 days. The young are fed on insects and insect larvae by both parents and they leave the nest when they are 2 weeks old.

GREY-WINGED BLACKBIRD
Turdus boulboul

The Grey-winged Blackbird is widespread from western Pakistan across India to southern China and the northern parts of south-eastern Asia. It reaches a length of 30 cm. The female is ash brown in colour and has a russet spot on her wings. This blackbird inhabits mountain woodland at heights from 1 800—3 000 m, but in winter it descends into the valleys to live at an altitude of about 1 000 m. At this time, it is also seen in parks and gardens. In spring the male perches high up on rocks or on dead branches in the treetops, and makes a flute-like call. The nest is built in bushes or trees, or in crevices in the rocks, and is made of grass stalks, roots, moss and various other plant materials. The female lays 3—5 spotted eggs and sits on them for 2 weeks. The chicks are fed by both parents, on insects, worms, spiders and molluscs. When they are 14 days old, and still unable to fly, the young jump out of the nest and hide among the stones and bushes. These blackbirds feed on insects and worms, and also on berries and soft juicy fruit.

GÜLDENSTÄDT'S or RED-BELLIED REDSTART
Phoenicurus erythrogaster

Güldenstädt's or the Red-Bellied Redstart is a typical denizen of high mountains from the Caucasus to the Himalayas, where it occurs at heights of 3 600—5 200 m. This is remarkable in a bird only 16 cm long. In winter it flies to lower-lying land, but never below 1 500 m. The female is predominantly a pale buff colour below, and fawn above, with a rufous tail. In summer this redstart lives in grassy, bush-covered localities, but in winter it settles in the valleys, beside the mountain streams. It builds its nest

Grey-winged Blackbird

Güldenstädt's or Red-bellied Redstart

in hollows in rock walls or among boulders. It is made of roots, stalks, moss and leaves, and is lined with feathers and hair. The female lays 3—6 eggs and sits on them for 14 days. When the clutch hatches out, both parents are kept busy, feeding the ever-hungry nestlings on flies, beetles, butterflies and spiders, often catching the prey on the wing. Adult birds also eat berries. The young leave the nest after 2 weeks, but the adults continue to feed them for another 10 days.

SPOTTED FORKTAIL
Enicurus maculatus

The Spotted Forktail, which reaches a total length of 25 cm, has a long, forked tail. It is native to the extensive mountain areas stretching from Pakistan to the northern part of south-eastern Asia. It is found at heights from 600—3 000 m, on the banks of moun-

Spotted Forktail

Hume's Wheatear

White-crested Laughing-thrush

tain streams. Here the forktail scurries among the stones searching for insects, spiders and small molluscs. It builds its nest in rocky crevices near streams, or in tree hollows. The nest is made of stalks, roots and moss, and the nesting cup is lined with animal hair. The female lays 3—6 eggs and sits on them for 2 weeks. The young are fed by both parents on insects and insect larvae.

HUME'S WHEATEAR
Oenanthe alboniger

Hume's Wheatear is found on arid rocky sites and in mountain valleys in the northern regions of southern Asia. It is about 16 cm long. This bird lives on the banks of streams and rivulets. Here it runs swiftly on the ground, hopping from stone to stone or sometimes perches on a high spot to survey its surroundings. It preys on beetles, butterflies, spiders and other invertebrates. Its nest is situated in a heap of stones or in a rocky crevice, and is made of roots, stalks and leaves. It is lined with feathers and animal hair. The clutch of 3—5 eggs is incubated for 14 days, mainly by the female. The young are fed by both parents.

WHITE-CRESTED LAUGHING-THRUSH
Garrulax leucolophus

The White-crested Laughing-thrush inhabits the Himalayas, its range extending to southern China, Taiwan, Malaysia and western Sumatra. It reaches a length of 30 cm and both sexes are alike in colour. This bird frequents mountain areas up to a height of 2 000 m, preferring wooded slopes with dense undergrowth, or bamboo thickets, from which its loud, laughing voice is often heard. Outside the nesting season it travels in flocks of 12—40, frequently visiting gardens and parks in villages or on the outskirts of towns. It forages on the ground, turning leaves over with its beak in its search for insects, spiders and molluscs. It sometimes catches small lizards, and takes young birds from their nests. It also feeds on berries. The nest is built among bushes. It is made from dry grass stalks and bamboo leaves, and is lined with fine moss and animal hair. The female usually lays 3 greenish-blue eggs. Both partners share in incubation for about 14 days, the female sitting on the clutch for longer periods. The chicks are fed by both parents.

SILVER-EARED MESIA
Leiothrix argentauris

The Silver-eared Mesia is found from the Himalayas
to southern China and across Burma, Laos and Vietnam to Sumatra. It is about 16 cm long. The female
differs from the male in that the underside of the tail
feathers is yellowish-olive, and her plumage is less
bright. This species frequents dense evergreen forests
up to a height of 2 100 m, though it is usually found
at about 1 000 m, wandering through the bushes and
treetops in search of insects, insect larvae, spiders,
seeds and berries. Outside the nesting season, it
roams in flocks of 30 or more, sometimes joining
groups of other birds, such as laughing-thrushes. In
the courtship period, the male perches on top of a tall
bush, singing loudly. The female builds her nest near
the bottom of a thick bush, while the male brings
building material for her. The nest is made from
roots, grass stalks, various fibres, hair and feathers.
The female lays 3—4 eggs and sits on them for
13—14 days, being relieved by the male only for brief
periods. The young are fed by both parents.

TIBETAN or GREY-BACKED SHRIKE
Lanius tephronotus

The Tibetan or Grey-backed Shrike ranges from the
Himalayas to western and southern China, and occurs
almost all over south-eastern Asia. It is a large species, reaching a length of about 25 cm. Both sexes are
alike in colour. It prefers bush-covered slopes and upland plateaux at heights of 2 700—4 500 m, descending in winter to lower-lying, more sheltered situations.
It usually lives in pairs, although some shrikes become
solitary outside the nesting season. This sturdily built
shrike preys mainly on large insects, often in flight,
but it also takes lizards, frogs, the young of small
birds and small mammals such as mice. Like other
shrikes, this species impales surplus food on sharp
thorns in the bushes, and stores it until times of shortage. In the breeding season, the male perches on the
branch of a tall bush imitating the songs of other
birds. The nest is usually situated 2—3 m above the
ground in a tall thorn bush. It is made of dry twigs,
roots and grass stalks, and the nesting cup is lined
with feathers and animal hair. The female lays 3—5
spotted eggs and incubates them for 15 days, occasionally being relieved by the male. Both parents feed

Silver-eared Mesia

Tibetan or Grey-backed Shrike

Green-backed Tit

137

the young on the nest for 19 days, and for a further
14 weeks after fledging.

GREEN-BACKED TIT
Parus monticolus

The Green-backed Tit inhabits an area stretching
from the Himalayas to western China, Burma and
north-western Vietnam. It is found on wooded slopes,
and in parks and gardens in mountainous areas from
1 200—3 600 m. It is about 13 cm long. In winter,
this tit lives at lower altitudes, occurring in flocks
which often include other species of small birds. In
spring, when the mountain trees and bushes are in
bloom, the Green-backed Tit builds its nest in a hole
in a tree or in a crevice in the rocks. It is made of
pieces of moss and lichen, and is lined with feathers
and animal hair. The female lays 5—8 whitish eggs
with russet spots at daily intervals, and incubates
them for 13—14 days, while the male brings caterpillars, termites and various other insects to her. When
the young hatch, their parents arc kept busy trying to
satisfy the ever-hungry chicks, for just one family can
consume several hundred small hairless caterpillars
every day. After about 18 days, the young tits leave
the nest, but their parents continue to feed them for
a further 10 days. After fledging the families roam together in the neighbourhood of their nests.

SULTAN TIT
Melanochlora sultanea

The Sultan Tit is distributed from the Himalayas to
southern China and Hainan, and across south-eastern
Asia to Sumatra. This beautiful tit is 20 cm long. The
female has a dark olive mantle and a yellowish-olive
throat. The Sultan Tit abounds on wooded slopes
where there are tall trees. It usually occurs at heights
from 300—1 000 m, ascending to 1 800 m in some
places. Outside the nesting season it wanders in small
flocks in the treetops. There the birds climb nimbly
along the branches and twigs, and even on the leaves,
often head downwards, in search of insects and insect
larvae. They also eat seeds and soft berries. The birds
keep in touch with one another, every now and then
calling with a pleasant trilling phrase, repeated three
times. In spring each pair occupies a territory and defends it against intruders. The nest is built in a hole
vacated by woodpeckers, usually about 3 m above the

Sultan Tit

Rock Nuthatch

Fire-tailed Sunbird

ground in a tree. The female lays 3—6 rust-spotted eggs and incubates them for 2 weeks. The male brings food to her, and later both adults feed their young on insects, insect larvae and spiders.

ROCK NUTHATCH
Sitta neumayer

In rocky limestone areas of central Asia, one can sometimes come across narrow-necked flasks attached to high vertical rock walls. In the sunshine, their surfaces seem to glitter with emeralds and rubies. What artist could have placed these jewel-studded works of art at such a height? It is the 19 cm-long Rock Nuthatch. This bird makes a chattering noise as it flies and skilfully squeezes into its bottle-shaped nest. This extraordinary construction is made of wet mud, brought from a nearby stream. The birds mix the mud with saliva and make layer after layer, smoothing them down with their beaks. They finish by making the elongated entrance, and sometimes strengthen the walls with the shining, metallic wing-cases of various beetles including rare leaf beetles, which are greatly prized by entomologists. The female lays 3—5 eggs with delicate reddish-brown spots, and incubates them for 13 days. The young are fed by both parents. Nuthatches hunt among the stones and vegetation beside the streams, preying on insects, especially locusts, grasshoppers and beetles. At the age of about 3 weeks the chicks leave their unusual nest and roam through the countryside.

FIRE-TAILED SUNBIRD
Aethopyga ignicauda

The Fire-tailed Sunbird is distributed from the Himalayas to south-western China and northern Burma. The male reaches a length of 15 cm and has a long, narrow tail. The female is 10 cm long, and is olive-coloured, with a yellow abdomen and rump. This agile bird is found in thin coniferous woodland with thick undergrowth, or in rhododendron bushes and junipers at heights from 3 000—4 000 m. In winter the birds descend to the valleys, to altitudes between 1 200—2 300 m. These birds feed on nectar and on small insects. The bowl-shaped nest is built among branches. It is made of moss, stalks and feathers, and lichens growing nearby are used to camouf-

Brown Bullfinch

Greater Racket-tailed Drongo

lage it. The female lays 2—3 russet-spotted eggs. The young hatch out after 3 weeks and are fed by both parents.

BROWN BULLFINCH
Pyrrhula nipalensis

The Brown Bullfinch is found in the Himalayas and the high mountains stretching to southern China,

Yellow-billed Blue-pie

northern Burma and northern Vietnam. It is about 17 cm long. The female resembles the male, but she is predominantly yellowish in colour instead of red. In summer the birds live at heights from 2 100—3 000 m, ascending to 4 000 m in Sikkim. They live in coniferous or oak woodland, and among dense rhododendron bushes. They seek their food in trees and bushes, eating seeds, berries, buds and nectar. The female makes a nest of twigs which she breaks off, and she lines the nesting cup with lichens and animal hair. The male usually accompanies his mate or perches nearby, singing. The female lays 3—5 eggs covered with fine rust-coloured dots, and she sits on them for about 13 days. The male sometimes takes over for short periods. The young are fed on insects and insect larvae, and leave the nest after 14 days, after which they remain with their parents. Family flocks roam through the countryside near their homes, flying in winter to lower altitudes and sheltered valleys, though they always stay above 1 500 m.

GREATER RACKET-TAILED DRONGO
Dicrurus paradiseus

The Greater Racket-tailed Drongo is widespread throughout south and south-east Asia to the Greater Sunda Islands. It reaches a length of 35 cm, but a large part of this is taken up by the tail. It is found in evergreen forests and bamboo jungles ranging from the lowlands to mountainous areas at a height of about 1 500 m. It is often seen in tea plantations, where it seeks its food in tall treetops. It feeds on caterpillars, beetles, large mantises, butterflies and swarming termites. It can even catch dragonflies, and it occasionally pounces on small geckos among the branches. It wanders through the crowns of large trees, such as coral and silk-cotton trees, and sips nectar from their beautiful red blossoms. In the nesting season the birds make loud, melodious, whistling calls, usually early in the morning and before nightfall. Each pair builds a nest 5 cm deep and 15 cm across, made of thin twigs and strips of bark. It is often interlaced with cobwebs which are also used to attach the nest to a forked twig at the end of a branch, 5—15 m above the ground. Nests can also be found in hollows in huge palm leaves. The female usually lays 3 spotted eggs, varying in colour. Both partners share in incubation for 16 days and both rear the young.

YELLOW-BILLED BLUE-PIE
Urocissa flavirostris

The Yellow-billed Blue-pie inhabits mixed pine and oak forests and vast rhododendron groves from the Himalayas to south-western China and north-western Vietnam. This colourful bird is about 66 cm long including its very long tail. It is a resident of mountainous areas at heights from 2 100—3 300 m, descending in winter to a height of about 1 000 m. Outside the nesting season it wanders in groups, often with other species of birds of similar size. The flocks fly from tree to tree, pecking at berries and fruit in the treetops, or catching insects and small lizards such as geckos. They also prey on treefrogs and sometimes plunder the nests of small birds, taking both eggs and nestlings. They collect fallen berries on the ground, moving with a curious hopping gait, and with their tail feathers held well away from the ground. They sometimes cause considerable damage to crops in orchards and plantations. A flock of blue-pies has been seen to join vultures in feeding on the carcass of a large deer. Nesting usually takes place in May and June, at which time the birds make strange creaking sounds. Each pair makes a shallow nest of twigs 5—6 m above the ground in the top of a low oak tree or in a rhododendron bush. The nesting cup is lined

with roots and fine leaves. The female lays 3—4 pale cream eggs with reddish-brown spots. The clutch is incubated by both parents for about 17 days, and they feed the young mainly on insects.

TOKAY
Gekko gecko

The Tokay reaches a length of 40 cm, and is one of the largest species of gecko. Its range of distribution covers the whole of south-east Asia. It lives among the trees as well as upon rocks and it is found in houses in large cities such as Singapore. It obtains its name from the call it makes, and the Malaysians believe it brings luck to any house where it settles. During the day, the Tokay lies hidden in a rocky crevice, or in a crack in a wall. It comes out to hunt in the evening, and its barking voice can be heard at a distance of more than 100 metres. After nightfall, this gecko scurries swiftly over rocky walls or buildings, gripping with its broad fingertips which have layers of scale-like suckers on the underside. These adhesive pads enable it to climb up smooth surfaces, and even walk upside down. This species feeds on insects, smaller geckos, young mice and nestlings. Its sharp teeth can also be used in self-defence, but the gecko often falls prey to owls, snakes and small carnivores. The female lays her eggs always in the same place, perhaps in a rocky cleft or in a crack in the bark of a tree. There are several clutches a year, each comprising two eggs. The eggs have parchment-like membranes, which harden quickly in the air. The young take a very long time to hatch, sometimes as long as 200 days. After hatching, they seek their own territories, usually occupying them for life. Tokays have fragile tails which readily break off when grasped by an attacker. This can save the lizard's life and the tail soon grows again.

ASIATIC LEAF-TOED GECKO
Hemidactylus frenatus

The Asiatic Leaf-toed Gecko is a small species, only 6—8 cm long. It is found throughout south-east Asia, extending to Australia. It is one of the most abundant species, living on rocks and trees, and being found in great numbers in houses in villages and towns. During the day, it hides in cracks in walls and ceilings, under the bark of trees, in rocky clefts, and various other dark spots. It comes out after nightfall to run briskly over walls and ceilings, catching small insects. It mainly catches gnats, which makes this lizard extremely useful. It also preys on flies and small moths. The female lays clutches of 2 eggs, always in the same place. They are soft to begin with, but soon harden in the air. The young hatch after 2 months.

INDIAN PYTHON
Python molurus

The Indian Python can attain a length of 8 m. It inhabits mountain primary forests and jungles of Burma, southern China, Vietnam, Malaysia and Java. It is very abundant in the mountains of Vietnam, and it can often be bought in the markets from local snake hunters. These specimens, which are usually about 3 m long, are caught in winter, when the snakes hibernate in caves. The hunters collect the sleeping snakes, tie them up and carry them in baskets to the market place. They pierce the jaws of the pythons and fasten them together with bamboo ropes to prevent the snakes from biting. Pythons have highly palatable flesh, and their skins are used to manufacture a variety of objects. The female lays several dozen leathery-shelled eggs in a secluded place, either in a large hole in a tree or in thick bamboo undergrowth, and coils

Tokay

her body around the clutch. In this way she both protects the eggs and keeps them warm on chilly nights. She is able to increase her body temperature by moving her muscles. The young hatch after about 80 days. The female remains with the clutch throughout this time without taking any food, and only occasionally drinking from a nearby water-hole. The Indian Python feeds mainly on small mammals and birds, young snakes preying on mice and young rats.

ASIAN LAND SALAMANDER
Onychodactylus japonicus

The Asian land salamander *Onychodactylus japonicus* is native to Japan. It reaches a length of 18 cm and breathes through its skin, for it has no lungs. It lives beneath stones on the banks of mountain streams at a height over 1 000 m. Its diet consists of small insects, worms and molluscs. Early in spring, the males crawl from their underground shelters into still water, and are followed a few days later by the females. The eggs are laid in special sacs which the females attach to stones or aquatic plants. The larvae have curved horny claws which are sometimes retained in the adults.

GIANT SALAMANDER
Megalobatrachus japonicus

The Giant Salamander, which reaches a length of 150 cm, is the largest of the amphibians. It is indigenous to mountainous localities on the Japanese islands of Hondo and Kyushu. It frequents mountain rivers and streams at heights from 300—1 000 m, and spends its entire life in water. It passes the day among the tangle of roots beneath a bank or among large stones. At night it leaves its shelter to hunt insects, molluscs, worms, crustaceans, fishes and amphibians. In the egg-laying period, salamanders travel upstream in search of faster-flowing, better oxygenated water. Here, usually in August or September, the female lays two strings of yellowish eggs having a diameter of about 5 mm. Each gelatinous string is 14—18 m long and comprises 400—500 eggs, which are fertilized in the water. The larvae hatch within about 2 months and undergo metamorphosis at 3 years of age when they reach a length of 20 cm.

Another large species of salamander — *Megalobatrachus davidianus* — lives in the Chinese mountains.

BUTTERFLY
Danaus tytia

The beautiful diurnal butterfly *Danaus tytia* has a wingspan of over 10 cm. It is native to the slopes of the high mountains of India, Assam and the Himalayan foothills, where it occurs at a height of about 2 500 m. It flies during the day, often alighting on the blossoms of trees and bushes.

APOLLO BUTTERFLY
Parnassius autocrator

The apollo butterfly *Parnassius autocrator* inhabits the Himalayas and the Pamirs. It is a rare and beautiful butterfly with a wingspan averaging 6 cm. At the present time, it seems to have disappeared from the Himalayas, but it is still common in Afghanistan, where it frequents mountain meadows at heights from 3 300—3 500 m. On sunny summer days, it can be seen flying from flower to flower. It has excellent powers of flight and is not easily caught. Its caterpillars feed on various species of fumitories.

SWALLOWTAIL
Atrophaneura horishanus

The swallowtail *Atrophaneura horishanus* is a beautiful diurnal butterfly with a wingspan of up to 10 cm. It is distributed in eastern China and Taiwan, where it frequents mountain meadows at heights of over 2 000 m. In June, it glides above the grassy slopes often alighting on the flowers and sucking nectar with its long proboscis.

BUTTERFLY
Bhutanitis lidderdalei

The butterfly *Bhutanitis lidderdalei* is one of the most beautiful species of butterfly. This denizen of the high mountains of Bhutan, northern India, Burma and Szechwan has a wingspan of about 10 cm. It frequents wooded slopes, flying among the crowns of trees, and rarely descending to the ground.

Butterfly
Danaus tytia

Apollo butterfly
Parnassius autocrator

Swallowtail
Atrophaneura horishanus

Butterfly Bhutanitis lidderdalei

DESERTS
AND STEPPES

_____

Deserts cover vast areas of central Asia from east of the Caspian Sea, across the Gobi Desert to the Khingan Mountains. The Gobi Desert is the largest, followed by the Kara Kum and Takla Makan, and by the Thar Desert in north-western India. The barren Rub 'al Khali, also called the Empty Quarter, in Saudi Arabia, is another huge desert. Many deserts are sandy, while others are covered with stones and pebbles. The deserts of Asia are not continuous, in many places being interrupted by tracts of steppe with a cover of grass or bushes. Saline patches have formed in some areas, their surfaces glittering with silvery crystals of salt. Some have no vegetation, but others produce a few stunted plants and shrubs capable of growing in the salty soil. The largest such region is found in the Thar Desert, which is less arid than the Asian deserts. Many plants and thorny bushes abound here, and after the rains, even grass appears. At this time, herdsmen arrive with huge flocks of sheep, cattle or camels, to allow the animals to graze on the lush pastureland. In years when there is a rather heavier rainfall, the local inhabitants sometimes grow millet and harvest wheat. The Thar Desert is crossed from north to south-east by the river Luni, which in some parts forms extensive, often very shallow salt lagoons. Salt is obtained from them by evaporation. The river gradually vanishes in the sand, but the water flows underground and can be reached by means of wells. However, the water is extremely salty. In the south, the low-lying Thar Desert blends into unproductive salt marshes.

The vegetation of the salty localities comprises many plants of the genus *Suaeda,* the stems of which have a frosted appearance. *Ipomoea biloba,* another salt-loving plant, with very long, thread-like stalks, thrives on the sandy plains. This beautiful, red-flowered plant grows in large formations, its long roots penetrating the sand, and curtailing movement of the dunes.

The trailing caper shrubs of the genus *Capparis,* characterized by semisucculent leaves and large whitish-violet blossoms which open in the early morning, are abundant in sandy and stony deserts, and on rocks and steppes. The long, thorny stalks virtually creep out, and in spring, they are densely covered with heart-shaped buds. These are collected and pickled in vinegar, to be used as a seasoning. Several varieties of caper shrubs occur in the deserts of central Asia, and in some areas they form thick carpets on the sand.

146

Low, thorny shrubs of the genus *Alhagi* cover large areas in some places. *Alhagi camelorum,* with small leaves and red flowers in spring, is most common in the sandy deserts and steppes of western and central Asia. Locally it is called camel grass, for it is the staple food of camels, the only domestic animals that are able to manage the large, prickly stems. The leaves of this species are also eaten by sheep. On young shrubs the thorns are soft, but they soon dry and harden, and to walk through a thicket of *Alhagi* is an excrutiating experience. The prickles hurt the feet and the tips often break off and remain in the skin where they can cause severe inflammation.

Saxaul trees of the genus *Haloxylon* are characteristic of the deserts and steppes of western and central Asia, and they occur in several species. These trees are low, their slender trunks with tail-like branches attaining an average height of about 2 m. They usually grow in groups, and their shady crowns shelter many animals from the scorching sun. An unusual plant, *Cistanche flava,* often grows beside these trees. Its succulent stem, sometimes up to 40 cm tall, sprouts from the hot sand and in spring bears many small flowers, buff to reddish-mauve in colour. Numerous desert rodents obtain the liquid necessary for their survival from its stems. How did this plant find moisture in the desiccated soil? The explanation is simple. *Cistanche* grows directly from the roots of *Haloxylon.* In the same way it is parasitic on *Ephedra,* drawing water from their roots. Several species of the genus *Salicornia,* characterized by segmented, apparently leafless succulent stems, are found in salty localities throughout the Asiatic deserts.

The deserts produce green vegetation only for a brief period in spring. In summer there is no escape from the burning rays of the sun and the hot winds which swirl the tiny grains of sand. Even in the shade, the temperature may rise to more than 50° C, and the sand is like embers to the touch. At this time all the vegetation becomes withered and scorched, but life remains in the harsh arid environment. The desert comes to life mainly after nightfall, when the temperature falls and the animals emerge to look for food. Nevertheless, even in the most intense heat, some creatures can be encountered during the daytime, especially in spring. For example there is a surprising number of beetles of the genus *Julodis.* The vegetation of the deserts of central Asia, especially the verges of the Kara Kum, feature many shrubs of the genus *Calligonum,* reaching to heights in excess of 2 m, and the beetles can be seen on their branches among the stunted leaves. In spring, the branches are covered with the *Julodis variolaris* or black *Capnodis excisa,* and the shrubs look as if they are thick with berries.

The most commonly found insects are locusts of many species, sizes and shapes, and the low, often modest vegetation teems with their flattened bodies. When they are disturbed, hundreds may rise together into the air. Some can be quite frightening, for they make loud rattling sounds to threaten their enemies. However, many insectivores are not to be scared away; and swarming locusts are often covered by a huge pinkish cloud of hundreds, or even thousands, of rose-coloured starlings attracted by the prospect of easy prey. Locusts and various other insects are caught by agamas or by speedy racerunner lizards of the genus *Eremias,* and highly predatory mantises often take them and devour the soft parts of their bodies.

The sun sets around 6 p.m. and it is soon dark. Then a grumbling, droning sound begins to be heard. It is made by large black beetles, the sacred scarabs, which cruise in the air like miniature aircraft. They leave their shelters in the sand to search for animal excrement, from which to mould balls of dung to feed their larvae. On the edges of the deserts, where sheep graze, these beetles follow the flocks in huge numbers. Camels also supply welcome material for these characteristic desert beetles. Many other beetles traverse the fine blown sand, leaving clear traces in their wake. These include large numbers of black beetles of the genus *Pimelia,* which hide during the day in dug-out holes obscured by the sand.

Evening forays are also undertaken by small pythons of the genus *Eryx,* which prey on tiny lizards. They lie in wait, buried in the sand, or slowly crawl on the surface in search of hidden lizards or agamas. After dark, the flowering bushes become surrounded by small whitish cockchafers and many tiny moths. A lighted torch will at first illuminate only the magic outlines of the sand dunes or the bizarrely shaped shrubs, and nothing else of particular consequence. However, when the torch is played more closely on the surroundings, an amazing phenomenon presents itself. A bright green light blinks on the ground, disappears and shines again. When

the torch is switched off, the strange lantern vanishes. When it is switched on once more, instead of the expected green light, a bright red flame appears. This desert mystery is really quite easily accounted for. The green light comes from a small desert spider, the luminous abdomen of which shines in the light of the torch. The flaming red lights are reflected from the eyes of the gecko *Teratoscincus scincus,* as it forages for insects. These handsome lizards are especially abundant in some localities. Nocturnal mammals also come out to feed. The Long-eared Hedgehog roams the dunes, its excellent sense of smell enabling it to find insects and small vertebrates. Foxes also prowl the desert sands. At dawn, the nocturnal animals return to their shelters, leaving behind them a variety of tracks leading in all directions. At first these can be followed to the animals' burrows and holes, but the wind soon covers all traces of life with a veil of sand.

New prints soon appear on the blown sand, this time belonging to the diurnal denizens of the desert, such as slow land tortoises, or predatory desert monitors.

Of the many dangers lurking on the edges of deserts and in the steppes, not least is the danger of poisonous snakes. One of the most fearsome snakes is the Levantine Viper, *Vipera lebetina,* which blends with its background, and can easily be inadvertently stepped upon. If a serum is not administered in time, the venom of this snake can prove lethal. The cobra is also highly venomous, but unlike the viper, it usually flees in good time and fatal accidents are rare.

The deserts often blend gradually into steppes which in turn merge into mountain areas going up to heights in excess of 2 000 m. On the steppes, shepherds can often be encountered with large herds of sheep, stunted cattle, donkeys or camels — dromedaries in the western regions and Bactrian camels in the east. In some places, particularly in the hills, huge herds of domestic animals have devastated the landscape.

The plateaux of central Asia are often covered by patches of giant fennel of the genus *Ferula,* which reaches a height of 3 m. These interesting plants have thickened, beet-like roots, once used in medicines. Some species of this fennel are inhabited by large, beautiful beetles of the genera *Melosia* and *Agapanthia,* the larvae developing on the roots of the plants. In the same localities 150 or so species of the genus *Astragalus* are to be found. The dense, prickly shrubs are up to 1 m high, and usually have a domed, cushion-like habit. They have tiny white flowers. In the hills, the profusely flowering thorny shrubs of the genus *Paliurus* grow to a height of over 2 m. Purple beetles of the species *Purpuricenus budensis* and many other species of beetles live on these shrubs, and a variety of flies are attracted by the strong scent. Metallic green beetles of the genus *Sphenoptera* live on the tallest plants. In the steppes, lowlands and hills, there are many tortoises such as *Testudo graeca* or *Testudo horsfieldii*. Grassland steppes are also the favourite haunts of locusts and grasshoppers. Vast numbers of insectivorous birds, particularly various species of buntings and shrikes, are attracted by such insects, for they provide an abundance of food for their large families. Locusts are preyed upon in particular by rose-coloured starlings, common and blue-cheeked bee-eaters and rollers. Flycatchers are common in the hills where they fly close to the ground in the dark, skilfully catching insects in their open beaks. The steppes also provide suitable habitats for many rodents, and these are followed by birds of prey, chiefly pallid harriers, tawny eagles and upland buzzards.

Tall foxtail lilies of the genus *Eremurus*, which grow to a height of 2 m, are found on the steppes of the high plateaux of western and central Asia and in the Himalayas. From their dense tufts of broad, elongated leaves grow tall stems with thick clusters of star-shaped flowers, coloured white, yellow or pink. *Eremurus algae* bears yellow flowers, while *Eremurus comosus* produces white blossoms. Poppies of the genus *Papaver* abound in the hills, where their bright red flowers shine among the greenery. *Papaver pavoninum* and the Long-Leaded Poppy *(Papaver dubium)* are among the most common species. In spring the central Asiatic steppes are dotted with hundreds of beautiful tulips. Some are yellow, others are orange, and the species *Tulipa davidiana* has flowers with longitudinal violet streaks.

Tamarisks of the genus *Tamarix* are characteristic shrubs of the valleys, steppes and deserts of central Asia. Their blossoms attract all kinds of insects, particularly flies and wasps, and small birds build their nests on the branches. Mulberry, one of the most widespread trees or shrubs of Asia, is found beside streams throughout the steppes. The Black Mulberry *(Morus nigra)* is native to western Asia, while the White Mulberry *(Morus alba)* originates from China.

The sweet fruits of these trees provide food for many species of birds, and they also enrich the diet of the native population. Mulberries became famous because their leaves are the food of caterpillars of the Silkworm Moth *(Bombyx mori)*. Since ancient times, the Chinese have made the finest silk from the cocoons of this extraordinary moth. Nowadays the species is raised throughout the world, in localities where the climate favours mulberries.

This survey would not be complete without mention of those Asiatic plants which have become important cultivated crops. Sugar Cane *(Saccharum officinarum)* is a perennial grass indigenous to Bengal. It has densely segmented stalks which reach a height of 2—5 m and measure up to 5 cm across. Sugar cane is nowadays found in the humid tropics throughout the world. It grows in vast plantations, mostly in India, Java and Cuba. The juicy pulp contains 12—18 per cent of sugar, and the remaining molasses are used to make rum.

Common Millet *(Panicum miliaceum)*, the stalks of which are about 1 m tall, is native to the steppes of central Asia. In China, it was grown as a crop as early as 3 000 B. C., and in many countries, such as India, it now ranks among the most important crops.

The Tea plant *(Thea sinensis)*, which is now cultivated in hillside plantations, has also become world famous. Many varieties are grown in tropical and subtropical Asia. Tea is a bush or low tree, originating probably from a wild ancestor, *Thea assamica*, a shrub native to Assam. Tea has been cultivated in China since the 4th century, and in Japan since the 9th century. Extensive plantations were introduced in India, Sri Lanka, Java and Sumatra in the 19th century, and cultivation has spread to other countries of Asia. In Europe, tea has been known since the turn of the 16th century, and in the early 18th century, it made its appearance in America.

In gardens or on rocky sites beside streams on the steppes, trees or shrubs with twisted, crooked trunks are often found. In spring, their branches are covered with magnificent, shining red flowers, visible over large distances. This plant, the Pomegranate *(Punica granatum)* bears red, apple-like fruit. The tree itself is of no particular importance, but its fruit once had symbolic significance, and often appeared in the embroidery of local women. The pomegranate is nowadays cultivated as an ornamental plant.

LONG-EARED HEDGEHOG
Hemiechinus megalotis

The Long-eared Hedgehog is native to central Asia
and Afghanistan. This rodent is about 30 cm long,
and has conspicuously large ears. It inhabits mainly
sandy, bush-covered deserts, and can also be found in
arid steppes. During the day, the hedgehog shelters in
burrows up to 50 cm deep which it digs under bushes.
The burrow has only a single entrance, and it is used
throughout the year. This hedgehog has a fixed territory in which it hunts. It wakes up after nightfall and
starts to search for locusts and grasshoppers, beetles,
small lizards and snakes. It occasionally eats carrion.
After a gestation period of about 40 days, the female
bears 1—5 young in the burrow which she lines with
grass and leaves. The young hedgehogs are born
blind, and with soft quills beneath their skin. However, the quills sometimes begin to appear on the surface of the body as soon as 20 minutes after birth. The
young open their eyes after 2 weeks.

STEPPE PIKA
Ochotona pusilla

The Steppe Pika inhabits rocky terrain in the steppes
and semideserts of Kazakhstan. It reaches a length of
about 15 cm, and has a characteristically short tail. Pikas are related to hares, and are sometimes called piping hares on account of their shrill call. They often
live in large colonies, digging burrows up to 40 cm
deep, terminated by a nesting chamber usually with
two exits. The nest is lined with fine grass and the
leaves of bushes. Each female has her small range
within the colony, where the male joins her in the
breeding season. After a gestation period of 20—24
days, the female bears 7—13 young in her nesting
chamber, which is lined with hair and leaves. The
young are born naked and blind, opening their eyes
after 8—12 days. They leave the burrow when they

are 2 weeks old and move around near the entrance.
They are suckled for 20—22 days, after which time
they become independent. Although there are 3—4
litters a year, colonies do not become overcrowded,
for these animals form 90 per cent of the diet of birds
of prey and carnivores in the area.
Pikas feed on grass and leaves, often storing large
leaves and plants beneath stones. Each animal may
store up to 8 kg of dried material.
Other species of pika live in the Himalayas and in
central and eastern Asia.

JAPANESE HARE
Lepus brachyurus

The Japanese Hare reaches a total length of 55 cm including the tail, which measures 5 cm. It has relatively short ears. Its range of distribution covers eastern
Asia and the Japanese islands. In the north-eastern region of the island of Hondo and in the mountains of
Japan, the hare assumes a white coloration. This species is found in a variety of locations, mainly in
steppes, fields and semideserts, but also on the edges
of forests and in gardens. During the day it hides in
bushes or tall grass, where it has a regular sleeping
place. It begins to forage at dusk, eating grass, leaves,
shoots and growing crops, and it also feeds on seeds
and bark, especially in winter. After a gestation period of 42 days, the female gives birth to 1—5 young,
usually twice or three times a year. The young are
born fully-furred and with their eyes open. They are
suckled for 19—21 days, but they begin to take plant
food after the first week. They can fend for themselves
after 5 weeks.

BOBAK
Marmota bobac

The Bobak is distributed in steppes, semideserts and

Long-eared Hedgehog

Steppe Pika

Japanese Hare

Bobak

cultivated areas of central Asia. It reaches a total length of 65 cm, of which 12 cm is made up of the tail. This rodent lives in colonies, each pair digging out a burrow with its entrance either on a slope, or on an elevated hillock, to protect the nest from flooding by rainwater. Marmots sleep in their burrows at night, foraging by day and resting in the midday heat. They feed on roots, grass, leafy plants, fruits and seeds, and they also take insects and sometimes birds' eggs found on the ground. After a gestation period of 40—42 days, the female gives birth to as many as 12 young in her underground chamber. The young are blind, opening their eyes after 3 weeks, and leaving the burrow when they are about 5 weeks old. They are suckled for 8—10 weeks. In autumn, marmots carry dry grass to the burrow, and stuff the entrance with it. Then they hibernate in family groups of as many as 15, made up of pairs and their offspring. The body temperature of the hibernating marmots falls to around 4—8°C, and the animals take only 2—3 breaths each minute, compared with the normal 16.

GOLDEN HAMSTER
Mesocricetus auratus

The Golden Hamster is well-known as a domestic pet. It is about 18 cm long including the tail which measures just over 1 cm. Females are stouter than males. The Golden Hamster is native to the eastern Caucasus. It was first described as early as 1839, but became domesticated only after 1930 when a single female and her twelve young were brought to Europe from Syria. In some languages, it is consequently

called the Syrian hamster. Its habitats are in steppes and semideserts, and here it digs out burrows in which to sleep during the day. It forages after nightfall, feeding on seeds, green plant material, insects, molluscs and worms. Gestation lasts only 16 days and the female bears 4—8, or occasionally up to 12 young in a burrow, as much as 1 m deep. Breeding takes place at all times of the year. The young are born blind and naked, opening their eyes after 14 days and suckling for 3 weeks. They become independent after 7—8 weeks. This rodent can live to be 3—4 years old, but its usual life span is 2 years. It is active even in winter, often coming out during the day.

DWARF HAMSTER
Phodopus sungorus

The Dwarf Hamster is a small species reaching a length of only 6—11 cm including the 1 cm-long tail. This rodent dwells in the steppes and semideserts of Mongolia and in adjacent regions of northern Chi-

Golden Hamster

Dwarf Hamster

na and Siberia. It often settles among colonies of pikas. In winter its fur becomes white and very dense. The Dwarf Hamster excavates burrows where it shelters during the day. It comes out to forage before nightfall, feeding on seeds, berries, grass, insects and worms. The female gives birth to 4—6 young after a gestation period of 17—18 days. Young hamsters are born naked and blind, opening their eyes when they are 12 days old. They are independent after 3 weeks.

COMMON MONGOOSE or NEWARA
Herpestes edwardsi

The Common Mongoose, or Newara, is distributed from the southern and eastern regions of the Arabian Peninsula to India, Sri Lanka and Burma. It reaches a length of about 85 cm, but 35 cm of this is taken up by the tail. It frequents steppes and sandy and rocky terrain where bushes provide good cover, and it may occasionally be encountered in thin woodland. The Common Mongoose is well-known for its ability to kill cobras and other venomous snakes. It moves so quickly that it is able to seize a snake behind the head without the snake being able to strike back. The fight is usually brief and the mongoose quickly kills and eats its opponent. However, snakes do not form the bulk of its diet. It feeds mainly on small rodents, birds' eggs, and also on insects and molluscs. The

mongoose lives in small groups or in families, in rocky clefts or underground holes. The gestation period lasts 55—65 days and a litter consists of 2—4 young, born sparsely covered with fur. The young open their eyes after 8—12 days. There are 2 litters each year.

PRZEWALSKI'S HORSE
Equus przewalskii

Przewalski's Horse is the only surviving species of wild horse. It is a sturdy animal reaching a height of up to 145 cm at the shoulder and a length of 200 cm. Its tail is 90 cm long. An adult stallion is heavier than a mare, and weighs up to 350 kg. In summer the coat is very short, but in winter it becomes longer and more closely-packed, to protect the horse from the intense cold in its homeland. This wild horse was discovered in 1879 by the explorer N. M. Przewalski on his travels across Dzungaria to Mongolia. His discovery caused a sensation because the wild ancestors of horses were thought to be extinct. Przewalski brought back a skull and a hide of this wild horse, and these served as a basis for the description of the new species. The first live wild horses were taken to Europe in 1899 and safely reached the town of Askania Nova in the Ukraine. In 1901, the famous Hamburg-based zoo-owner Carl Hagenbeck, sponsored a Mongolian expedition which returned with both stallions and mares which formed the basis for Przewalski's Horse to be bred in captivity. The native Mongolian hunters knew of Przewalski's Horse before its official discovery and killed it for its meat which was believed to have miraculous powers. This was one of the reasons why its numbers fell rapidly. Furthermore, with the spread of pasturelands, the wild horses were obliged to move to more arid parts of the steppes, where they did not flourish, for they needed to drink regularly.

Common Mongoose or Newara

Przewalski's Horse

The few horses now remaining in the wild are distributed in a large part of south-western Mongolia, particularly in the Gobi Desert and in the Altai Mountains. In autumn, they migrate several hundred kilometres south to eastern Tien Shan, returning to the north in spring. They are most usually found in steppes and semideserts of the upland plateaux at heights from 700—1 800 m, or even as high as 2 500 m. Here the horses feed on dry grass or plants for most of the year, or browse on the leaves and branches of stunted bushes. They regularly visit water-holes after sunset and cannot survive for more than 3—4 days without water. Przewalski's Horse is sociable, living in small groups made up of 5—15 mares with their foals, and led by a strong stallion. On sensing danger the horses flee, with the leader in the rear to protect the mares and foals. If a herd is attacked by a pack of wolves, the horses form a circle with the young in the middle. The adult horses stand with their heads to the centre, and kick out with their hind legs to frighten the wolves away. These horses can run at a speed of up to 70 km per hour over short distances and at over 50 km per hour when running steadily. In summer, they walk in Indian file, each horse using its tail to remove irritating flies from the head of the following horse. During snowstorms, the herd forms a circle with the foals in the middle and the adults keep them warm with their bodies and by breathing on them. In spring the female gives birth to a single young after a gestation period of about 336 days. The foal weighs about 45 kg and is able to stand up after only a few hours. It begins to follow its mother after 2—3 days.

ASIATIC WILD ASS
Equus hemionus

The Asiatic Wild Ass occurs in five subspecies, but its systematic classification has not been standardized. The Onager (*E. h. onager*) is found in northern Iran and in an adjacent region of Russia. It stands about 120 cm high at the shoulder. It is similar to the Khur (*E. h. khur*) of north-western India, Pakistan and south-eastern Iran. The Central Asiatic Wild Ass or Kulan (*E. h. kulan*) lives in Turkmenia and Kazakhstan, and the Dzeggetai (*E. h. hemionus*), which is about 115 cm tall, inhabits Mongolia. The Kiang (*E.

Asiatic Wild Ass

h. kiang) is 150 cm tall and is the largest subspecies. It lives on the plateaux of Tibet at heights from 4 000—5 000 m. Some zoologists classify the Kiang as an independent species, while others distinguish another subspecies of the Asiatic Wild Ass, the Syrian Wild Ass (*E.h. hemippus*).

The Asiatic Wild Ass has been exterminated in many places and faces extinction in others. Large herds once roamed the steppes, but herdsmen with their horses and sheep penetrated the heart of the vast steppes and forced the wild asses to leave their native habitats. Hunters armed with modern weapons easily shot these cautious animals, and in eastern Asia they were eagerly sought because their meat and blood were believed to have miraculous powers. Wild asses are also preyed upon by wolves, which can catch weak or diseased animals, though they never attack a herd. During the day, wild asses feed on grass, salt-loving plants and the leaves of bushes. In the evening, they drink at water-holes though they can survive without water for 2—3 days. At night, they seek shelter in the valleys or among thick bushes. Each group is made up of 6—10 individuals, led by an adult stallion, who always puts himself between the herd and its attackers to give the mares and foals time to make good their escape. In autumn, when the Asiatic wild asses travel further afield in search of food, smaller groups merge in herds of as many as several hundred. The rutting season begins in early summer, the stallions fighting each other for possession of the herds of mares, and often engaging in fierce combat. After a gestation period of 330—360 days, the mare bears a single young, which is able to walk within a few hours. It weighs 25—30 kg at birth. At first it hides in the undergrowth, but after a week it is able to run with the herd. It begins to graze when it is a month old, but continues to be suckled for almost 12 months. The females mature after 3 years, the males after 4—5 years. In the wild, the Asiatic Wild Ass lives to be about 15 years old, but it can live for up to 30 years in captivity. When under attack, a herd can run at 60 km per hour. Over short distances the wild ass can run as fast as 72 km per hour, and its average speed of about 40 km per hour can be maintained for long distances.

BACTRIAN CAMEL
Camelus bactrianus

The Bactrian Camel is a large animal, reaching a height of 230 cm at the shoulders, a length of 420 cm and a weight of up to 600 kg. This two-humped

Bactrian Camel

camel now occurs in a domesticated form, bred in Asia and in parts of south-eastern Europe. Wild camels can still be found in western Mongolia, the adjacent region of China, and in eastern areas of central Asia, though they are not often encountered. They frequent upland plateaux, even in the high mountains, their long, dense winter coats enabling them to withstand the cold and the freezing winds. In spring, they lose their hair rapidly, in large tufts. Camels live ideally at heights from 1 500—2 000 m, in places where there is good grazing. However, the herdsmen have forced them to leave most of these localities. In summer the camels are scattered all over the vast steppes, but they form small groups of up to 20 in winter, and seek situations protected from the wind. Each herd is led by the strongest male. In the rutting season, which usually takes place in February, the male gathers several females together, and fights to retain them. Rival males bite each other and often inflict serious wounds. They also try to frighten each other by inflating the soft palate, which looks like a large bubble coming out of the mouth. During the rutting season, even in captivity, camels can be a danger to man. After a gestation period of 385—406 days, the female gives birth to a single young in which the two humps are already evident. The young camel is capable of following its mother within a few hours. The Bactrian Camel is a ruminant, feeding predominantly on grass, other green plants, twigs and fruits. It is a very undemanding animal. It can go without water for as long as 12 days, after which it is able to drink over 135 litres of water in ten minutes. When it travels as part of a caravan, it has to be given water at least once every three days. The camel is an ambling animal, lifting both legs on one side at the same time, and therefore rocking from side to side. Hearing is its best-developed sense, though it has a good sense of smell, which enables it to detect water from a considerable distance.

Camels have an unpleasant characteristic — they have a tendency to spit the foul-smelling, chewed contents of their stomach at their adversaries, while at the same time producing loud roaring sounds. In the wild, adult camels have almost no enemies, though a young camel which has strayed can be taken by wolves.

Bactrian camels are much used as beasts of burden, for they can carry a weight of over 250 kg. A caravan travels slowly, covering only 30 km a day, but it remains the best means of transportation in many desert

areas. In central Asia, the Bactrian Camel was domesticated five to six thousand years ago.

BLACKBUCK
Antilope cervicapra

The Blackbuck is widely distributed in India, in the area south of the Himalayas and between the rivers Indus and Ganges, where it inhabits steppes and thin woodland. A subspecies which once lived in northern India is now extinct. The adult male stands up to 85 cm tall at the shoulders, is 120 cm long and weighs

Indian White-backed Vulture

Asian King Vulture

about 40 kg. The female is smaller and more slenderly built. The Blackbuck lives in groups of 15—50, although several hundred of these antelopes may occasionally gather. Old bulls often lead a secluded life, seeking mates only in the rutting season, when they fight for small harems of females. Following a gestation period averaging 180 days, the female bears a single young or occasionally twins. The young antelope is suckled for 5 months, but eats green vegetable food of all kinds from the second month. Blackbucks graze in the morning and evening, resting during the day in the shade of bushes and trees. At one time cheetahs were trained to hunt these antelopes.

SAIGA
Saiga tatarica

The Saiga is native to Kazakhstan, Turkestan, southern Siberia, western Mongolia and north-western China. This antelope was once found throughout Europe and across Asia as far as the Lena River. At present, it is most abundant in Kazakhstan, where its population is estimated at 2 million. Huge herds of saigas migrate from the northern regions to the south in autumn, returning again in spring. Over 100 000 animals often make the trip together, leaving widespread destruction in their wake. An adult saiga stands 60—80 cm high at the shoulder, is 120—170 cm long, and weighs 36—63 kg. Females are less robust and normally lack horns, though occasionally they are seen in older animals. The males have ringed, pointed horns, up to 35 cm long. Saigas grow a short coat in summer and a long, dense one in winter.

The Saiga dwells in treeless steppes and semideserts, or sometimes even in fields, if it is not disturbed. It roams throughout the countryside, covering great distances, and swimming across rivers in its path. The enormous herds are led by females and their offspring, followed by smaller groups of males. At the onset of the breeding season, in November, the males engage in duels, the winner taking 5—50 females. The males take no food in the rut and become very thin. In the late spring, after a gestation period of 140—150 days, the female bears one, or sometimes two young, in a place covered by tall grass where the young can be hidden from such predators as wolves, foxes and large eagles. The young are suckled for 2—2.5 months. Females mature after 8 months, and males after 20 months. Saigas running from danger can reach a speed of 70—80 km per hour, and over long distances run at 60 km per hour. They feed on a great variety of plants, even those which are poisonous to other animals, and when the vegetation is moist and succulent, saigas can survive for long periods without drinking.

INDIAN WHITE-BACKED VULTURE
Gyps bengalensis

The Indian White-backed Vulture is distributed throughout India, extending to south-western China and western areas of south-east Asia. It is about 90 cm long. It is found on open grassland sparsely covered with bushes and trees, up to a height of 2 500 m, but it occurs in the mountains only in the more southerly areas of its distribution. It usually lives in small groups, often together with other species of vulture. These birds cruise for hours over the countryside, often following grazing herds of sheep or cattle for days, and patiently waiting for an animal to die. Then they fly in from far and wide, as if at some signal, and as many as 50 vultures can assemble at the carcass. They devour everything except the largest bones. Then, gorged, they sit around on the ground or in a nearby tree, until the food is sufficiently digested for them to fly to their regular roosting sites. Trees used regularly by vultures can be completely destroyed by their excrement. Breeding takes place between October and February, but most clutches are found in December. The vultures remain together at this time, forming colonies of 20—40 pairs on a small site along the edge of a forest. Each pair builds a nest in a large tree, often in a fig, mango or rosewood tree. The nest is made of branches and dry sticks, the nesting cup being lined with hair and scraps of hide. The female usually lays a single white egg with a greenish sheen, sometimes covered with reddish-brown spots. Both partners incubate the egg for about 47 days, and both feed the chick on regurgitated food. At the age of 80 days the young vulture is fully fledged and takes its first flight.

ASIAN KING VULTURE
Sarcogyps calvus

The Asian King Vulture inhabits India and the western half of south-east Asia. This sturdy raptor is about

Pallid Harrier

80 cm long, and has a bald head and leathery flaps on each side of its head. It lives in the open grassland of semideserts, and can also be seen in thin woodland. In Nepal and the Himalayas it occurs at a height of 2 000 m on open slopes. It usually lives in pairs, and it is unusual to see more than one or two of these birds together at a carcass. These vultures also visit the outskirts of large cities where they feed on domestic rubbish. This helps prevent the spread of dangerous infections by removing the remains of decaying flesh. In the courtship period both sexes produce sounds not unlike the whinnying of horses. Nesting takes place between December and April. The nest is built by both partners 9—12 m above the ground, usually in the top of a fig or mango tree, and often near a town or village. The massive flat nest is made of branches and is lined with leaves and grass stalks. In

Pied Harrier

semideserts the nest is only about 2—3 m above the ground, usually on a mesquite of the genus *Prosopis*. The female lays a single white egg which is incubated by both partners for 45 days, and the chick is fed by both parents.

Shikra

PALLID HARRIER
Circus macrourus

The Pallid Harrier, which is about 50 cm long, is found in the steppes and open scrubland of central Asia and of the temperate Asiatic zone as far as Lake Baikal. It also occurs in eastern Europe. This beautiful raptor has a wingspan of 105 cm. The female is predominantly dark brown above, and has a reddish head and underparts. In central Asia the Pallid Harrier occurs up to a height of 1 350 m, in the Altai Mountains it is seen up to 2 300 m, and in the Pamirs it ascends to 2 750 m. It is one of the most common raptors of the steppes. When hunting, it flies slowly just above the ground in search of prey which it usually seizes on the ground. It catches small rodents and insectivores, birds, and large locusts and grasshoppers. It occasionally also catches lizards and frogs. In autumn, the Pallid Harrier leaves for India, Sri Lanka and Burma, but the western population overwinters in tropical Africa. The birds form pairs before returning to their nesting grounds in April. They build a nest in a shallow depression lined with dry grass stalks. The female lays 3—6 white eggs, lightly dotted with brown, and incubates them for 29 days while the male feeds her. He also brings food for the young, partly plucking it before passing it to his mate to divide into portions for the young. When they have grown up a little, she also begins to hunt. Young harriers leave the nest when they are 40—45 days old.

PIED HARRIER
Circus melanoleucus

The Pied Harrier occurs in eastern Asia, ranging from the Amur River and the Trans-Baikal region to Manchuria, Korea and northern China. It reaches a length of 46 cm. The crown, wingtips, tail and mantle of the female are brown and her cheeks are whitish. The Pied Harrier frequents fields and bush-covered semi-arid grasslands. In the north it is also a bird of swamps and tundras. In autumn it travels south and south-east to India, Sri Lanka, south-eastern Asia, Borneo and the Philippines. In its winter grounds it often lives in rice paddies, returning to its homeland in early May. Pairs make a scrape on the ground, about 5 cm deep and 40 cm wide, and line it with dry stalks and leaves. The female lays 4—5 green eggs spotted with brown, and sits on them for about

28 days. If she perishes during this time, the male incubates the abandoned clutch. He feeds the female throughout incubation and brings food for the whole family for the first few days after the young hatch out. Later both parents feed the young. The Pied Harrier moves in graceful glides 1—2 m above the ground as it searches for rodents, birds, reptiles and frogs. This species seizes its prey on the ground and is unable to catch birds on the wing.

SHIKRA
Accipiter badius

The Shikra is native to central Asia, India and south-eastern Asia. It is a resident bird except around the Caspian Sea and in south-eastern Europe, from where it migrates to the Arabian peninsula to spend the winter. It is a small raptor, the male reaching a length of 30 cm and a wingspan of 60 cm, while the female is 36 cm long and spans up to 80 cm. The Shikra is common in deserts, settling in oases, or in cultivated areas containing groves of trees. In the Himalayas, it is found at a height of 1 500 m. It often travels 3—4 km from the nest in search of prey, scouring the sandy deserts for lizards, small rodents and birds up to the size of a partridge. It also catches locusts and grasshoppers. The nest is built in trees, on a branch 5—12 m above the ground. It is made of branches and is lined with hair. The pair sometimes occupies the nest of other birds, and sometimes the birds build 2—3 nests near each other, finally settling in one of them. In May the female lays 3—4 whitish eggs with a greenish sheen, sometimes with pale grey spots. She incubates the clutch for 4 weeks while her partner brings prey for her to a regular site near the nest. He also feeds the whole family for a week after the young have hatched, joined later by the female, who divides the kill for the chicks. Young shikras leave the nest when they are a month old, and begin to learn to fly and hunt.

UPLAND BUZZARD
Buteo hemilasius

The Upland Buzzard ranges from central and eastern Asia, south as far as Tibet. This raptor, which is about 70 cm long, is predominantly resident, but some birds wander to south-eastern Asia in winter. The Upland

Upland Buzzard

Buzzard mainly inhabits rocky steppes and mountainous areas. In the Altai Mountains it occurs up to a height of 2 300 m; in Tibet it ascends to 4 500 m and frequents the same situations as the pika or piping hare *Ochotona melanostoma*, which forms the bulk of its diet. The Upland Buzzard also catches steppe and desert rodents, fowl-like birds, reptiles, frogs and toads. It also eats large insects, especially locusts and grasshoppers. It flies low over the ground when hunting, and seizes its prey on the ground. Each pair has two nests in an extensive territory, occupying them in

Tawny Eagle

Long-legged Buzzard

Laggar Falcon

alternate seasons. The nest is built on an inaccessible rock, or in low-lying steppes, in a tree. The construction is made of branches and is lined with moss, leaves and hair. The 2—4 yellowish, russet-spotted eggs are incubated by both partners for a month. The female sits on the nest longer than the male. At first she feeds the young on flesh brought by the male, but later he joins her in feeding the chicks. The young buzzards leave the nest when they are 45 days old, but their parents continue to feed them for another 4 weeks while they are learning to hunt.

TAWNY EAGLE
Aquila rapax

The Tawny Eagle is widespread in steppes and deserts of central, eastern and southern Asia. This bird of prey is about 75 cm long. It is resident in the southern regions of its distribution, migrating from northerly areas to India for the winter, while the western populations travel to north-east Africa. It prefers flat, low-lying country, covered by grass and bushes, but it also visits sandy deserts. It forms pairs on arrival at its nesting grounds. At this time, eagles engage in graceful courtship displays high in the sky. They make nests of branches and dry grass, 1—2 m above the ground among scattered trees, usually poplars or saxaul trees. Sometimes they nest on the ground on sandy islands or in a heap of old hay or straw. The construction measures 70—120 cm across. The female lays 1—3 white eggs with brown spots and incubates them for 45 days. When she is hungry, she has to find her own food, for the male does not feed her. However, the young are fed by both parents, although it is the female who divides the prey and passes it to the eaglets. These are fully fledged after 55 days and take their first short flight within the next few days. The Tawny Eagle preys on rodents, young fowl-like birds and on lizards and snakes. The prey is always caught on the ground.

LONG-LEGGED BUZZARD
Buteo rufinus

The Long-legged Buzzard is distributed from Asia Minor across central Asia and Afghanistan to Mongolia. It is 51—66 cm long, and occurs in two colour forms, of which the rufous one is more common than the dark brown. Populations of northern areas over-winter in India or in north-east Africa. The Long-legged Buzzard inhabits steppes, deserts and mountainous areas, ascending to a height of up to 4 000 m in the Pamirs and in Kashmir. Rodents form the mainstay of the Long-Legged Buzzard's diet, and in seasons when they are especially plentiful nests are found sometimes only 50 m from each other. Nesting takes place in March and April, the nests being built on a cliff ledge or in trees. In the Kara Kum Desert they are built on the low saxaul trees. The nest is made of twigs and dry grass. The female usually lays 2 eggs, off-white in colour, with dark brown spots, but if food is plentiful, there may be up to 5 eggs. Incuba-tion is undertaken by both partners for 29 days, and the young are fed in the same manner as the Upland Buzzard. This buzzard preys on ground squirrels, young hedgehogs, lizards and snakes, including venom-ous vipers, frogs and young birds.

LAGGAR FALCON
Falco jugger

The Laggar Falcon is indigenous to the whole of In-dia, Afghanistan, Assam and Burma. This raptor is 45 cm long and has a wingspan of 120 cm. When it hunts it swoops through the air and dives headlong down with its wings pressed close to its body as it aims at its prey. It is one of the fastest birds, and rare-ly misses its target. It mainly hunts myna birds, dron-gos and pigeons, though it also catches rodents, liz-ards, locusts and grasshoppers. It occasionally visits villages and takes domestic poultry. The nesting sea-son varies with the area of distribution, taking place in January in southern India, in February in northern India, and in March and April in the Himalayas. The pair builds its nest on a rocky ledge, on tall city min-arets or in a tree, 10—15 m above the ground. The male collects the twigs, grass and leaves, and the fe-male constructs the nest. She lays 2—5 eggs, usually pink with red or reddish-brown spots. She sits on them for 29 days and is occasionally relieved by the male. The newly hatched young are covered with thick down. For the first few days, the male brings food for the young, passing it to his mate to the ac-companiment of loud cries. She plucks the prey and divides it into suitable pieces. The male brings food straight to the young when they are 14 days old. At the age of 35—40 days, young falcons leave the nest to take their first flights.

Lesser Kestrel

LESSER KESTREL
Falco naumanni

The Lesser Kestrel, which measures about 30 cm, in-habits central Asia, eastern and southern Europe, and north Africa. In autumn it leaves in large flocks for more southerly regions in India, the Arabian Peninsu-la and Africa. In its homeland it is found in steppes and semideserts, and it often settles in cities, on min-arets, cathedrals and other high buildings. In the nest-ing season it lives in colonies of up to 200. Kestrels

Black Partridge

build their nests in holes under roofs, in wall crevices, cracks in stone bridges, tree hollows and clefts in rock walls. The female usually lays 5 white eggs, densely spotted with red. Incubation is carried out mainly by the female, though she is occasionally relieved by the male. The young hatch after 28 days and the female stays with them for the first few days, the male bringing food for the whole family. The female later joins him in hunting for prey. Young kestrels leave the nest at the age of 26—28 days, but their parents continue to feed them while they are learning to hunt. Their prey consists mainly of small rodents, lizards, locusts and grasshoppers.

BLACK PARTRIDGE
Francolinus francolinus

The Black Partridge is widespread from Asia Minor to Assam, and it is also found in Cyprus. The female is brownish with black spots and has a rusty stripe on her nape. She does not have spurs. The Black Partridge is a common bird which lives in groups of 3—6 for most of the year. In the breeding season it forms pairs which defend their own territories. Black partridges are found only in thick vegetation with bushes and tall grass for cover. Here the female makes a shallow depression, usually under a tamarind bush, and lines it with leaves and grass. She lays 4—10 eggs, yellowish-brown in colour, with dark specks. She incubates the clutch for about 22 days. The nesting season takes place between April and October. This fowl-like bird feeds on seeds, shoots, insects and other small invertebrates. It is preyed upon by raptors and small carnivores.

SEESEE PARTRIDGE
Ammoperdix griseogularis

The Seesee Partridge is distributed from the upper reaches of the Euphrates to north-western India. It reaches a length of about 25 cm, the female being smaller than the male and having brownish upper parts with lighter wavy lines and pale grey underparts with fine brown lines. The Seesee Partridge is an inhabitant of arid, stone-covered plains where there is little vegetation. Here it is found particularly on the hillsides, living in flocks of 20 or so. It forms pairs in the nesting season, and the female makes her nest on

Seesee Partridge

Daurian Partridge

Demoiselle Crane

a steep slope, usually among stones or clumps of low grass. The nest is a shallow depression lined with a few stalks of dry grass. She generally lays 8—12 eggs, though occasionally there may be as many as 16. They are normally pale yellow in colour. The female incubates the clutch for 21—23 days, and the young are reared by both their parents. At night and when they are resting, the female shelters them under her wings, and until they are able to fly, she regularly brings water for them in her beak. The adult partridges drink twice or three times a day. They feed on seeds, green plants, insects, insect larvae and spiders.

DAURIAN PARTRIDGE
Perdix dauuricae

The Daurian Partridge, a fowl-like bird about 30 cm long, is indigenous to central and eastern Asia, ranging to the Ussuri River region and to Mongolia. In the south, it reaches north-eastern Tibet. The female lacks the large abdominal patch. This partridge is a resident bird, settling in scrubland steppes, grassy mountain slopes up to a height of 3 000 m, and on the verges of forests. In winter the birds gather usually in flocks of 15—30, though occasionally there may be as many as 200 individuals. Partridges living in the mountains move down to the valleys in winter. They usually forage in the early morning and before nightfall, but even during the day they can be seen collecting seeds, and berries, or catching insects, spiders and molluscs. They also peck grass and green shoots. They roost at night among steep rocks. In spring the hen makes a shallow scrape in a clump of grass or under a bush, lining it with grass or fine leaves. There she lays 8—15 eggs, or sometimes more, and sits on them for 23—25 days. Throughout this time, the cock stays close at hand, guarding his mate. Both parents help the young to find food, and protect them from bad weather.

DEMOISELLE CRANE
Anthropoides virgo

The Demoiselle Crane is the smallest species of crane, but it is the most common. It can reach a length of 95 cm and a weight of 2.5 kg. It is widespread in the steppes of western and central Asia, extending in the east to Mongolia, and it also occurs in Africa and

Houbara Bustard

southern Spain. In winter the Asiatic populations migrate to India, Burma and to the Red Sea coast. In its native land the Demoiselle Crane inhabits river valleys and grassy or stony, bush-covered steppes and semideserts. In the breeding season, each pair occupies a territory of about one square kilometre, though in places where these birds are abundant, the distance between two nests may be a mere 200—300 m. These cranes use the same nesting sites for several years. They breed in April in the south, and from June onwards in southern Siberia. The nest is made on the ground, usually in the grass. It is just a shallow scrape sparsely lined with grass stalks or even without any lining at all. The female lays 2 spotted eggs and sits on them for 27—28 days, during which time the male occasionally relieves her. While one of the partners is on the nest the other patrols about 100 m round the nesting site, on the lookout for danger. The young, which can run as soon as they are dry, are cared for by both parents. When the brood has fledged, the families merge into large flocks of up to several hundred, and fly to their winter quarters. In spring they return in smaller groups of 10—30 birds, and soon form pairs. The Demoiselle Crane feeds mainly on insects, particularly locusts, and on green plants, seeds and berries.

Bengal Florican

White-tailed Lapwing

Greater Sand Plover

HOUBARA BUSTARD
Chlamydotis undulata

The Houbara Bustard has an area of distribution stretching from the northern regions of the Arabian Peninsula across Armenia to central Asia. The male reaches a length of about 65 cm and a weight of 2.3 kg, while the female is smaller and weighs only 1.4 kg. This bustard inhabits steppes, and desert areas with scattered vegetation. It also settles in cultivated fields. It is resident in southern areas, but birds of northern areas migrate to southern Arabia and Pakistan in winter. In the breeding season, in May and June, pairs of bustards defend their vast territories, nests usually being situated 5—10 km apart. Bustards do not build nests, the female merely making a shallow scrape beside a tuft of grass or under a bush. She lays 2—3 olive coloured eggs with grey spots and incubates them for 28 days. In sandy deserts, sand storms can completely cover the clutch, or even the sitting female. The male stays on guard during incubation but the brood is reared by the hen. She chases even large animals away, and has been known to attack men. She often pretends to be wounded, flapping her wings against the ground and limping as if she were lame. In this way she leads the intruder away from her family before suddenly taking to the air. When the young have fledged the small family flocks join the males. The Houbara Bustard is omnivorous, preying on locusts, grasshoppers and small lizards, and also feeding on seeds, capers and on a variety of green plants.

BENGAL FLORICAN
Eupodotis bengalensis

The Bengal Florican is a denizen of grassland steppes stretching from northern India to Burma. It reaches a length of 75 cm, the female being slightly smaller and rather paler in colour than the male. This species is solitary for most of the year, only occasionally gathering in groups of 4—8 birds. The nesting period is usually in March, but sometimes nesting also takes place during the summer months. At these times the male makes a jingling, metallic call. The female lays 2 olive green eggs with purple, brownish and greyish spots, in a depression lined with grass, and incubates the clutch for 30 days. The male remains with his mate only in the courtship period and takes no further care of the family. The Bengal Florican feeds mainly on plant material such as seeds, shoots, berries and grass, and to a lesser extent on insects, small lizards and snakes.

Black-bellied Sandgrouse

WHITE-TAILED LAPWING
Vanellus leucurus

The White-tailed Lapwing inhabits the grassland steppes of central Asia, in the south ranging as far as Pakistan. It attains a length of 28 cm. In autumn it migrates in large flocks of several hundred to northern India and to the Nile basin in Africa, occasionally wandering westwards to central and western Europe. In late March or April it returns to its nesting grounds situated near areas of swampland. The flocks then break up and form pairs, which build nests on the ground among tall grass, lining them sparsely with leaves and grass. The female lays 4 pear-shaped eggs densely covered with olive spots. Incubation is carried out by both partners for 24 days. Young lapwings stay in the nest for 2 days, and are then taken by their parents to the steppes, where food is abundant. Lapwings feed on locusts and other insects, and on spiders, worms and molluscs. At the age of 35 days, the young begin to fly and move around in small groups of 6—25 until the flocks set out on their journey south.

GREATER SAND PLOVER
Charadrius leschenaultii

The Greater Sand Plover, which is about 25 cm long, is found in central Asia, extending to Mongolia and north-western China. It frequents desert areas where there is only sparse vegetation, but where water, sometimes brackish, is always available. It catches insects on the plants, especially small locusts, and beetles such as beautiful leaf beetles of the genus *Sphenoptera*, which have a metallic sheen. It also preys on worms and small crustaceans in shallow water. The Greater Sand Plover makes its nest in a tuft of grass or among stones. The nest consists of a shallow depression thinly lined with grass stalks. The female lays 3—4 densely spotted eggs. She incubates them during the day and the male relieves her in the evening and at night. The chicks hatch after 25 days and leave the nest as soon as they are dry, to follow their parents to

sources of food. When they are 3 weeks old the young plovers are able to take short flights. In autumn, the birds merge into flocks and travel to their winter quarters on the coasts of east Africa, Madagascar, India, the Sunda Islands and Australia. This species has even been seen in Europe.

BLACK-BELLIED SANDGROUSE
Pterocles orientalis

The Black-bellied Sandgrouse lives in an area stretching from Iran across central Asia to western Mongolia. It attains a length of 35 cm and the female differs from the male in coloration. Her back, wings, head and breast are predominantly yellow with dense, black wavy lines, and her throat is yellowish in colour. This sandgrouse inhabits sandy areas of steppes

Collared Scops Owl

Mottled Wood Owl

In winter, sandgrouse may be seen even in north-western India.

COLLARED SCOPS OWL
Otus bakkamoena

The Collared Scops Owl is widespread throughout southern, south-eastern and eastern Asia and on the Sunda Islands and the Philippines. It reaches a length of 23 cm. Within this vast area of distribution there are 18 subspecies differing mostly in their coloration. The Collared Scops Owl frequents rocky, tree-covered slopes up to a height of 1 600 m, but it also lives along the edges of forests and even in city parks and gardens. It normally builds its nest in a tree hollow, usually in the trunk of a dead palm tree, but it may sometimes nest in a crevice in the wall of a building. The female lays 2—3 white eggs directly on the bottom of the hole and incubates them for 26 days. The young are fed on insects by both parents. The adults feed on small rodents and birds as well as on insects. The owls hunt their prey after nightfall.

MOTTLED WOOD OWL
Strix ocellata

The Mottled Wood Owl inhabits open, bush-covered localities and cultivated fields with scattered trees, in India and western Burma. It is about 50 cm long. It nests in hollows in large trees, or settles in church spires. The female lays 3—5 white eggs and sits on them for 28—30 days, during which time the male feeds her, hunting for prey after nightfall. For 10 days or so after the young have hatched, the female stays on the nest, the male bringing small rodents, bats, birds, amphibians and large insects for his family. When the female also begins to hunt, she stays on guard near the nest during the day. Young owls leave their nest after a month, and perch on nearby branches where their parents bring prey to them. They are able to fly when they are about 50 days old.

and deserts, but it can also be encountered in mountainous areas up to a height of 800 m. It lives in pairs in the breeding season. The female does not build a nest but lays 2—3 eggs in a shallow scrape on the ground. The eggs vary in colour, from reddish-yellow to reddish with brown and chestnut spots, sometimes blurred. Both partners incubate the clutch for 21—22 days and both care for their young, helping them to find seeds, green plant material and insects. The parents often carry water to their young over a distance of 50 kilometres or more. They store it in their feathers and the young suck it out. When the brood is fledged, the family joins up with other families to form large flocks which roam through the countryside.

SPOTTED LITTLE OWL
Athene brama

The Spotted Little Owl is native to India and the whole of south-eastern Asia. It reaches a length of

Spotted Little Owl

21 cm. It favours open localities and is often seen on church towers, on pagodas and beneath the roofs of buildings. It also lives among scattered trees in tea plantations. It dwells up to a height of 1 400 m in the mountains. During the day the Spotted Little Owl hides in hollows or among dense branches. In the courtship period, the male makes barking sounds. The nest is built by the female in a hole in a tree, in a crevice in a wall, or in a rocky cleft or cave. The clutch of 3—5 eggs is laid directly on the bottom of the hole and the female sits on it for 28 days. The young are fed on insects, small rodents and lizards by both parents. They leave the nest when they are a month old.

INDIAN ROLLER
Coracias benghalensis

The Indian Roller has an area of distribution ranging from the Middle East to south-western China and south-eastern Asia. It measures about 33 cm in length. It inhabits open country with bush cover, plantations and thin woodland, and parks and gardens in cities. In Nepal it is encountered at a height of 3 700 m. Rollers often perch on wires on the edges of roads, from where they survey their surroundings and keep a lookout for prey. They catch butterflies and termites on the wing, and fly down to the ground to peck up locusts or catch frogs and small reptiles. They occasionally kill small rodents or plunder the nests of small birds, eating both eggs and nestlings. In the courtship season, pairs of rollers engage in complicated aerial displays. The female lays 2—4 pure white eggs on a lining of fine grass stalks, in a hollow in a dead palm or other tree, usually 5—8 m above the ground. Both parents incubate the eggs for 17—19 days and share in the care of the young.

BLUE-CHEEKED BEE-EATER
Merops superciliosus

The Blue-cheeked Bee-eater, which is about 30 cm long, is widespread from the Middle East to central Asia and from India to south-eastern Asia. Outside the nesting season, the western populations migrate to tropical Africa, while the eastern populations overwinter in Sri Lanka, the Greater Sunda Islands and the Philippines. Bee-eaters travel in small groups, mainly in the early morning, resting during the day

Menetries's Warbler

Common Babbler

Isabelline Chat

Red-whiskered Bulbul

or foraging for food. They hunt on the wing, catching dragonflies, wasps, bees, and other insects, keeping watch for their prey from vantage points such as telegraph wires along the roads. In April or May the birds return to their native land and found nesting colonies. Each pair builds a tunnel, 1—2 m long, in a clay or sandy wall. The female lays 4—5 white eggs in a small chamber thinly lined with grass stalks, at the end of the corridor. Incubation is shared by both partners for 3 weeks. The young are fed on insects, which the parents catch in the surrounding steppes or deserts, often far from the nest.

BUSH WARBLER
Cettia diphone

The Bush Warbler is found in eastern Asia, the Japanese Islands and the Philippines. The male reaches a length of 16 cm, the female being about 2 cm smaller. This warbler frequents grassy localities where there are dense bushes, bush-covered mountain slopes, or sometimes sparse woodland above thick undergrowth. In the nesting season, the male makes a characteristic warbling call. The nest is built low down in thorn bushes or in tall grass. It is woven from grass stalks and leaves, and the walls are attached to twigs or tough stalks. It is about 10—13 cm tall and the nesting cup is up to 9 cm deep. The female lays 4—5 eggs and both partners take turns in incubating them for 13—14 days. They both feed their young, mainly on caterpillars and small flies. In autumn warblers migrate to their winter grounds, particularly to south-eastern China.

MENETRIES'S WARBLER
Sylvia mystacea

Menetries's Warbler, which measures about 13.5 cm, is found in southern regions of central Asia, from where it ranges across the southern shores of the Caspian Sea to Syria. It spends the winter months along the Nile, in Africa. In its native land this warbler lives in tamarisk shrubs along the banks of streams and rivers, and in saxaul trees, and in Armenia it is even found on rocky slopes up to a height of 1 350 m. After arriving at their nesting grounds, the males sing loudly from high branches. The nest is usually built in May, near the ground in a tamarisk shrub. The nest is

about 8.5 cm across and 8.5 cm high, the nesting cup
being 4.5 cm deep. It is woven from thin twigs and
grass stalks, with a lining of fine roots or horsehair.
The clutch of 4—5 whitish eggs with brown and grey
dots is incubated by both partners for 12—13 days.
The young are fed by both parents on insects and
spiders, and leave the nest after 12 days.

RED-WHISKERED BULBUL
Pycnonotus jocosus

The Red-whiskered Bulbul, which reaches a length of
about 20 cm, is widespread throughout south and
south-east Asia. It has also been introduced in Australia, in Mauritius and in the south of the United
States, everywhere successfully. It is very abundant in
its homeland, where it can often be seen in villages,
and in parks and gardens in large cities. It also occurs
in the mountains up to a height of 1 600 m, preferring
localities overgrown with bushes, in which it builds its
nest of roots, dry grass, leaves and wool, with a lining
of lichens. The nest is built by the female, but the
male brings the building material. The construction is
usually situated 1—2 m above the ground. The female lays 2 or occasionally 4 eggs, whitish in colour,
with red or reddish-brown spots. Both parents incubate the clutch for 12—14 days and both feed the
young. The chicks leave the nest after 12 days, but the
parents continue to bring food to them for some time.
After the fledging of the brood, bulbuls move in flocks
of about 30, but 2—3 more broods may often be
reared. The Red-whiskered Bulbul feeds on berries,
seeds, insects and spiders, and on sweet, overripe fruit
and nectar which it seeks in orchards and gardens.

COMMON BABBLER
Turdoides caudatus

The Common Babbler is about 23 cm long. This songbird is characterized by its long tail, and it has the appearance of a miniature pheasant. It inhabits dry,
open, scrub-covered country, cultivated land and gardens, from central Iran to Pakistan and throughout
India. In the Himalayas, it can be seen up to a height
of 2 100 m. It prefers tamarisk bushes, where the
pairs build deep bowl-shaped nests of small twigs,
leaves, moss and lichen. The female lays 3—5 greenish-blue eggs and sits on them, assisted by the male,

Black-headed Shrike

for 13—15 days. Both parents feed their young on insects, insect larvae and pieces of fruit. Adult birds also
eat seeds, berries and nectar from flowers, mostly of
the creeping caper shrub *Capparis aphylla*. The Common Babbler seeks its food mainly on the ground,
where it hops swiftly. It is also adept at climbing
along branches, but it is a poor flier. After the young
have fledged, babblers wander in small groups of
6—20, roosting in regular sites among the dense
branches of trees or tall bushes. The Common Babbler often plays host to a young cuckoo.

ISABELLINE CHAT
Oenanthe isabellina

The Isabelline Chat, which measures about 16.5 cm,
is an inhabitant of steppes, deserts and rocky areas
stretching from Syria across central Asia to Mongolia
and southwards to Pakistan. In the Pamir massif, it
can be encountered at a height of 6 000 m, although it
always nests below 4 000 m. This chat spends the
winter in Africa, along the Persian Gulf and in north-
western India, returning in March to its native land.
Soon after arrival, the nest is built by the female,
while the male stays near at hand. The relatively large
nest takes 5 days to construct. It is made of dry stalks

Red-rumped Swallow

Lichtenstein's Desert Finch

Desert Sparrow

ders. Young chats leave the nest after 2 weeks but are fed for a further 10 days.

BLACK-HEADED SHRIKE
Lanius schach

The Black-headed Shrike has an area of distribution covering all of southern, central and south-eastern Asia, the Sunda Islands, the Philippines and New Guinea. This bird is about 25 cm long. It inhabits extensive bush-covered localities, plantations, the edges of forests and mountain slopes, being found in the Himalayas up to a height of 3 400 m. Nesting takes place between February and June. Each pair builds a nest of twigs and grass stalks, lining it with fine roots or grass. The nest is situated 2—3 m above the ground, usually in a thorn bush or in an acacia. The female lays 2—4 eggs, pale green to yellowish in colour, with grey and brown spots, usually forming a circle at the broad end. The clutch is incubated mainly by the female, though she is occasionally relieved by the male. The young hatch after 2 weeks and are fed by both parents on insects, insect larvae, and morsels of small vertebrates, for 14 days in the nest and for another 10 days after fledging. Adult shrikes catch small geckos and sometimes take the nestlings of small birds. The Black-headed Shrike is a frequent host to various species of cuckoo, such as the Common Cuckoo *(Cuculus canorus),* the Common Hawk-Cuckoo *(Cuculus varius)* and the Jacobin Cuckoo *(Clamator jacobinus).*

and pieces of plant material, and is situated in an underground hole, an abandoned burrow or in a rocky crevice. It may be over 1 m deep, and at the end of a corridor up to 3 m long. The female lays 4—6 pale blue eggs, which she incubates for 13—14 days. The young are fed by both parents on insects and spi-

RED-RUMPED SWALLOW
Hirundo daurica

The Red-rumped Swallow, which is about 18 cm long, is widely distributed throughout southern, south-eastern and eastern Asia to Japan. It also occurs in southern Europe and Africa. It prefers rocky sites, being found in Nepal at heights up to 3 300 m, and it often settles on tall buildings in towns. It usually nests in pairs, which seldom form colonies. The nest is made of pellets of mud, which the swallow picks up off the ground. It plasters the pellets on cliffs, beneath overhanging rocks, on bridge supports and on the walls of buildings. The closed, hemispherical nest has a diameter of 20 cm, and is entered by a tunnel up to 19 cm long. The walls are 1.5 cm thick, the entrance

Rosy Pastor

is 4 cm wide and the whole nest can weigh over 1 kg.
The nesting cup is lined with fine moss and with
feathers, some of which can be up to 12 cm long. Both
partners take turns at sitting on the clutch of 3—7
eggs, and both feed the young on insects caught in
flight. Young swallows leave the nest when they are
20—21 days old.

LICHTENSTEIN'S DESERT FINCH
Rhodopechys obsoleta

Lichtenstein's Desert Finch is native to central Asia,
Iran and Afghanistan. It measures about 14.5 cm.
This bird also occurs in isolated nesting grounds in
northern China and central Mongolia. The female is
less subdued in appearance and lacks the dark band
stretching from the beak to the eye. This finch is
found in scrubland in steppes and deserts, near
streams or lakes. In the Turkmenian mountains it
lives up to a height of 2 000 m. It is a gregarious bird,
living in groups which form colonies in the nesting
season. A single bush is often occupied by several
pairs, the nests of fine twigs and lined with plant wool
being situated about 2 m above the ground. Pairs
form early in April. The clutch of 3—7 pale blue eggs
patterned with brown and black dots is incubated by
the female for 13—14 days. The young are fed by
both parents and leave the nest at the age of 2 weeks.
This finch feeds on seeds of various grasses, plants
and bushes.

DESERT SPARROW
Passer simplex

The Desert Sparrow inhabits desert areas of central
Asia. It is about 13.5 cm long and its light camouflage
coloration is perfect for life in a sandy environment. It
also takes up residence in towns and villages. The
nest is built by both partners, mostly on saxaul trees
and acacias, usually 2—3 m above the ground. It is
a spherical construction of dry stalks, thin twigs and
other plant materials, and the interior is thoroughly
lined with fine fibres and feathers. The female lays
5—6 whitish eggs with dense brown and greyish-purple spots. Incubation is carried out by both partners
for 13—14 days. At first the young are fed on insects,
progressing in due course to seeds, which form the
bulk of the Desert Sparrow's diet. These birds usually
consume seeds of the grass *Aristida pennata*.

ROSY PASTOR
Sturnus roseus

The Rosy Pastor is one of the best-known and also
one of the most useful birds of central Asia. The area
of distribution of this starling extends to Yugoslavia
and to the Persian Gulf. The Rosy Pastor is 21.5 cm
long and both sexes are alike in colour. It lives in
large groups in open country, especially on cultivated
land, and flies to desert areas in search of locusts,
which are its principal prey. There, large flocks looking from a distance like huge pink clouds, swoop
down to where the locusts are most abundant, and ef-
ficiently destroy them. These starlings also prey on
other kinds of insects and they enjoy ripe mulberries.
They spend the winter in India, Sri Lanka and around
the Persian Gulf, returning to their homeland in early
spring, to nest. The nesting colonies are made up of
hundreds, or sometimes thousands of birds. The nesting grounds are not fixed but vary with the distribution of locusts and the availability of rocks suitable
for nest-building sites. They also settle in tree hollows, in haystacks, or under the roofs of houses in the
villages. The nest is a flimsy heap of twigs, grass
stalks, straw and leaves, lined with roots and feathers.
The female lays 3—7 pale bluish-green eggs, and incubates them for 12—14 days. The young are fed by
both parents, and leave the nest after 14—19 days.

PANDER'S GROUND JAY
Podoces panderi

Pander's Ground Jay inhabits deserts of central Asia,
mainly the Karakum and Kyzylkum Deserts. It has

Starred Tortoise

Four-toed Tortoise

Whiskered Agama

Racerunner Eremias velox

Rock Lizard

long legs, well-adapted for running on the ground, where it seeks its food. It mainly catches locusts, other insects, spiders, and small lizards. In winter it also collects seeds. This bird seldom drinks, obtaining sufficient moisture from its prey. Nesting takes place in May, the birds building their bowl-shaped nests low down in saxaul trees or in various bushes. They are made from dry twigs, and are lined with grass. The female lays 4—5 eggs, highly variable in colour, but usually densely covered with violet-grey spots. She incubates the clutch for 16—19 days, while the male stays on guard and warns his mate when danger threatens. The young are fed on insects by both their parents.

STARRED TORTOISE
Testudo elegans

The beautiful Starred Tortoise is about 35 cm long and lives in bush-covered grassland in India and Sri Lanka. It has a knobbly, very high carapace. It is quite common, but it is expert at avoiding capture although it is hunted for its excellent meat. This tortoise is predominantly vegetarian, feeding on grass, green shoots, berries and fallen fruit. It also takes insects and occasionally eats fresh carrion. The female lays 6—8 hard-shelled eggs in a depression which she digs out in soft soil. The young hatch after 3 months and are immediately able to fend for themselves. In their early days they are heavily preyed upon by birds of prey and small carnivores.

FOUR-TOED TORTOISE
Testudo horsfieldii

The Four-toed Tortoise is found in the area around the Caspian Sea, in Turkestan, Iran and Afghanistan, and in the northern part of western Pakistan. Adult tortoises have a carapace about 30 cm long, and a characteristic feature of this species is the presence of four toes on its forelimbs. This tortoise frequents steppes and semideserts, being found in the mountains up to a height of 1 200 m. It feeds partly on green plants and various fruits, and partly on prey such as insects. It overwinters in burrows which it digs out for itself, occasionally occupying dens deserted by other animals. It comes out again in spring, and in May of June, the female lays 3—4 eggs. Young

females lay only 2 eggs. During the season each female lays 2—3 clutches. The eggs are about 5 cm long and the females bury them in the ground, afterwards stamping down the soil. The young tortoises hatch out after 80—110 days. The meat and eggs of this tortoise are eaten as a delicacy.

WHISKERED AGAMA
Phrynocephalus mystaceus

The Whiskered Agama, which is about 20 cm long, inhabits central Asia and Afghanistan. It is usually a uniform sand colour, but sometimes it has dark spots. This coloration enables it to merge with the sand of the deserts where it lives. It spends the winter in burrows 70—80 cm long, dug in the sand, and it comes out in early April. During the summer it buries itself in the sand at night by making jerky lateral movements of its body. It uses the same method to escape danger, managing to vanish in the sand within a few seconds. Only the vague outline of its body betrays the agama's presence. It remains motionless beneath the sand only fleeing when it is actually touched. Then it turns back frequently and attempts to scare the intruder away by taking up a threatening position, gaping and displaying broad protuberances around its mouth. This often deters birds or small carnivores. The diet of this agama consists of worms, beetles and their larvae, butterflies, locusts, spiders, and other small invertebrates. The reptile itself falls prey to monitors, boas, birds of prey and carnivores. There are two clutches a year, the first in late May or early June, and the second at the beginning of July. Each clutch contains 2—6 eggs, in membranous coats. The young hatch after 4—6 weeks and measure 3.5 cm.

RACERUNNER
Eremias velox

The racerunner *Eremias velox* is distributed in southern Mongolia, north-western China, central Asia, Afghanistan and Iran. It is up to 18 cm long. Young lizards have three brownish-black or black stripes on their backs, the middle one dividing on the neck. This species frequents sandy or partially sand-covered localities, and is sometimes found on stony terrain sparsely covered with grass and bushes. It also occurs

Gecko Teratoscincus scincus

in areas of clay and on the banks of rivers. In the mountains it may be seen up to a height of 1 500 m. It excavates vertical burrows, 15—20 m deep, or alternatively shelters in dens vacated by other animals, or beneath stones. Racerunners inhabiting northerly areas hibernate, leaving their shelters at some time between late February and April. At the end of May or in June, the female lays 4—12 eggs, and she deposits a second clutch in July. The young hatch after a month. These lizards feed on beetles, locusts, termites and spiders.
About 35 other species of this genus are found in the steppes and deserts of Asia.

ROCK LIZARD
Lacerta saxicola

The Rock Lizard is found in Crimea, Caucasia, Asia Minor, northern Iran and southern Turkmenia. It is about 15 cm long and variable in colour. It is usually seen on rocks and on stony sites, often beside rivers and streams. It also lives in trees along the edges of forests. In mountainous areas, it dwells at a height of 3 000 m. This lizard is expert at climbing vertical rocky walls and it can leap up to 40 cm. It hides away in crevices and clefts, and preys on insects and spiders. In late July or in August the female lays 2—4 eggs which hatch after 2 months. The young measure about 2.5 cm.

177

Schneider's Skink

Desert Monitor

Sand boa Eryx miliaris

Snake Taphrometopon lineolatum

Racer Coluber ravergieri

PANTHER GECKO
Eublepharis macularius

The Panther Gecko inhabits India, Afghanistan, Iran and Iraq, and is occasionally encountered in southern Turkmenia. It reaches a length of 30 cm. During the day it shelters beneath large stones or in the holes of various rodents. It hunts after nightfall, moving slowly over sandy or stony localities, seeking mainly insects. It catches beetles, locusts, grasshoppers, spiders and scorpions, and large individuals sometimes attack other small geckos. At night, it makes curious squealing sounds which betray its presence. The female usually lays 2 eggs under stones or in rocky crevices.

GECKO
Teratoscincus scincus

The gecko *Teratoscincus scincus* is widespread in central Asia, being most abundant in the Karakum Desert and Iran, and ranging to the Tien Shan Mountains, the Altai Mountains and the Pamirs. It reaches a length of 15 cm or more, and occurs in several subspecies. It is most often found where there are extensive areas of sand sparsely covered with grass and low bushes. During the day it shelters in burrows up to 80 cm long which it digs in the ground. In places where the soil is hard, it uses burrows abandoned by other animals. Fine particles of sand blown by the desert winds usually conceal the entrance. The lizard leaves its shelter after nightfall and slowly searches its territory for insects and spiders. It can easily be found in the dark, using a torch at eye level, for when the light strikes the eyes of this handsome lizard, they glitter like red jewels. It is quite easy to pick up this lizard, but care must be taken because the animal's skin is very delicate and can be damaged if it is handled roughly. The female usually lays 2 clutches of eggs, one immediately after the other, in June. Each clutch consists of 2 eggs, which measure about 18 mm in length.

Rat snake Elaphe dione

SCHNEIDER'S SKINK
Eumeces schneideri

Schneider's Skink ranges from central Asia to north-west India and north Africa, occurring in several sub-species throughout this vast territory. It reaches a length of over 20 cm. It is found in rocky situations near mountains, and in stony localities along the edges of forests. It also lives on the fringes of sandy deserts and on bush-covered mountain slopes up to a height of 1 500 m. It hides away underneath stones and in the holes of various animals, sometimes excavating its own burrows 40—100 cm long. In central Asia the female lays her eggs in late July, burying them in the ground. This skink feeds on locusts, beetles and other insects, large specimens also taking tiny lizards.

DESERT MONITOR
Varanus griseus

The Desert Monitor is a denizen of central and south-west Asia, and north Africa. It attains a length of 130 cm. It inhabits arid tracts of sandy or stony desert, living in burrows vacated by rodents or digging its own holes in the ground. It hunts in the morning or in the afternoon, resting during the heat of the day either in the shade or in the cool of its deep burrow. The Desert Monitor can run quickly and climb easily on bushes or slanting tree trunks. Here it seeks birds' nests, taking both eggs and nestlings. It also preys on lizards, snakes and rodents on the ground, and young monitors catch insects. The female lays 10—20 eggs which she buries in the sand beneath leaves or stones.

SAND BOA
Eryx miliaris

The sand boa *Eryx miliaris* inhabits the sandy deserts of central Asia. It grows to a length of 60 cm. This snake lives predominantly in sand and is adept at burying itself very quickly. After doing so, it remains motionless, but a shallow line in the sand above its body makes it easy to locate. In the late afternoon it comes out to hunt, crawling slowly across the sand in search of small rodents or lizards. Sometimes it lurks, buried in the sand, waiting for a lizard to pass by and

then adroitly seizing it in its jaws. It may occasionally attack a young monitor, and engage in a prolonged fight with it. The sand boa is surprisingly strong, considering its small size, but it is harmless to man, and when caught does not even try to bite. This snake spends the winter deep in a burrow vacated by rodents. It comes out in spring, and in late June or July the female bears 6—10 live young, about 13 cm long. The young are very active, feeding at first on insects or worms.

SNAKE
Taphrometopon lineolatum

The snake *Taphrometopon lineolatum* is found in Afghanistan, in the southern parts of central Asia and in Iran. It rarely reaches more than 1 m in length, and its body is extremely thin and conspicuous. Its eyes have a golden glitter. This slender snake lives in deserts, and it also frequents the edges of forests, fields, and mountains up to a height of 1 500 m. It often takes up residence in burrows abandoned by rodents. Adult snakes prey chiefly on lizards suffocating them in the coils of their bodies and biting them with venom fangs situated at the back of the upper jaw. The lizard dies within a few seconds. Men are seldom bitten by these snakes, and the venom causes only a slight swelling.

RACER
Coluber ravergieri

The racer *Coluber ravergieri* has a vast area of distribution, stretching across Palestine, Syria, Iraq,

Greater Indian Rat Snake

Iran, Georgia, Armenia, north-western India and Afghanistan. It seldom reaches more than 1 m in length. Entirely black specimens can be found in some parts. This racer lives in stony, semiarid regions, on rocky slopes and wooded foothills, in oases and high in the mountains. It feeds on small rodents, birds and lizards. In July the female lays 5—10 eggs, 4—5 cm long. The young hatch out at the end of August and in September. A newly hatched racer measures up to 20 cm.

RAT SNAKE
Elaphe dione

The rat snake *Elaphe dione* is distributed from northern Iran and Georgia across central Asia to southern Siberia, the Far East and China. This arboreal snake may occasionally reach a length in excess of 1 m. As well as living in the forests, it is also found on sandy steppes, reaching to the mountains. It is often seen in water, even seawater, and it consumes fishes although it feeds predominantly on rodents, and occasionally takes birds or birds' eggs in the trees. When it is disturbed or annoyed, it jerks the tip of its tail, making a faint rattlesnake-like sound among the leaves which discourages intruders. However, this snake is entirely harmless. The egg-laying season varies with the climatic conditions, but it usually takes place by mid-July, or in August in the Far East. The female lays 5—16 eggs, about 5 cm long. Incubation lasts a month, or somewhat longer in the north. The newly hatched young are about 20 cm long.

Indian Cobra

GREATER INDIAN RAT SNAKE
Ptyas mucosus

The Greater Indian Rat Snake is widely distributed throughout southern Asia, southern China, Taiwan, Afghanistan and Turkmenia. It can reach a length in excess of 2 m, and has large, conspicuous eyes. Some specimens are black with yellow abdomens. This rat snake prefers moist localities, usually near water, and is common in gardens, parks and scrubland. It feeds on frogs and tadpoles, and also on rodents. When it is disturbed, it can partly raise the front part of its body like a cobra. The female lays 8—16 eggs, up to 6.5 cm long. In China she lays them in May, while in India they are laid in July and August. When catching this snake in the wild, it is necessary to avoid its sharp teeth. Although this is a non-venomous species, its teeth can make tiny wounds that heal only slowly. This snake tries to avoid contact with man by taking refuge under a stone or in a hole. However, when its escape route is cut off, it often turns against its pursuer in an attempt to frighten him away.

INDIAN COBRA
Naja naja

The Indian Cobra is widely distributed and occurs in several subspecies. Cobras found in Sri Lanka and India have a characteristic pattern on their necks, varying somewhat from individual to individual. The cobras of southern India have a spectacle-like pattern on their hoods, while in the cobras of northern India, the pattern is made up of two simple spots. In eastern India, cobras have a single eye-like spot, sometimes with two small spots on each side. The subspecies in the most northerly and easterly areas do not have a pattern on their necks. The Indian Cobra has a vast area of distribution, being found in India, Sri Lanka, Iran, southern central Asia, southern China, Taiwan, Burma, Thailand, Laos, Vietnam, Indonesia and the Philippines. It reaches a length of 2 m, and varies in colour from uniform shades of fawn to a dark background striped or spotted with yellow or fawn.

The Indian Cobra remains always in localities where it can find plenty to eat, particularly rodents, so it is found along the edges of deserts, in semideserts and steppes, and it often takes up residence in caves, and in holes in the walls of human dwellings. It is nocturnal, emerging only after nightfall to seek its prey. Sometimes it takes domestic fowls or pigeons, and it has even been seen to catch fishes in shallow water. However the main part of its diet is composed of mice, rats and other rodents. An adult cobra can consume as many as 10 mice a day, after which it can go without food for as long as three weeks. It bites its prey with its venom fangs and then waits for it to die. In the case of small animals, this happens within a few seconds.

In the breeding season cobras live in pairs. The female digs out a hole with her snout and makes two entrances. In India, the female lays about 20 eggs, but in Turkmenia she lays only 10 or so. Both parents guard the eggs, one of them staying inside the corridor, near the entrance. The eggs have leathery shells and the young hatch out after 70—90 days. While the cobras are on guard, they are easily aroused, and readily rush from their shelter in an attempt to scare away any intruder. However, they never give chase as they are often said to do, and they usually try to avoid man.

Most cobras adapt easily to captivity, becoming passive and never being vicious. These qualities are made use of by native snake-charmers, who nevertheless remove the cobras' venom fangs, making these snakes as harmless as any domestic pet. However, the venom teeth have to be removed regularly and forgetfulness may prove fatal for snake-charmers who put their fingers into a cobra's mouth or as is their habit, offer the cobra a kiss. The picture of a snake-charmer sitting cross-legged with a raised cobra before him, dancing to the tune of his long flute, is well known. Less well-known is the fact that cobras are deaf and cannot hear the music. The sounds produced on the flute are superfluous, for the pendulum-like movement of the cobra's body is merely a response to the movement of the snake-charmer's flute. A tired cobra may sometimes attack but usually its gape is closed. Some snake-charmers display cobras which still possess their venom teeth, but such snakes, dazzled by the sun, do not attack even if they are taken out of their baskets with bare hands. In India and Sri Lanka, snake-charmers perform in almost every city. It is even possible to learn the art of snake-charming in special schools. Boys are able to enter the snake school when they are 6 years old. To begin with they learn how to handle the flute, how to sit, and so on. In due course they are taught to catch venomous snakes, identify them, treat them and pull out their fangs with a piece of cloth. After three years, the boys are ready to practise their bizarre craft. Anyone who is not afraid of snakes can become a snake-charmer — the cobra itself has little to learn.

Despite its non-aggressive nature, the Indian Cobra remains a highly dangerous snake because of its deadly venom. In India, where cobras are common, many people go barefoot and often step on a snake in the dark and get bitten when the cobra attacks in self-defence. If an anti-snakebite serum is not administered, the victim dies. The number of such deaths is 20 000 annually. The cobra venom enters the blood and very quickly stops the heart and the breathing. An adult cobra has the capacity to kill as many as 15 people. In comparison with vipers, the length of the venom fangs is diminutive, being only 3.5 mm long in adult specimens. A subspecies found in Malaysia and Vietnam can spit its venom over a distance of 2—3 m. The snake sprays the poison from the openings in its venom fangs while holding its mouth half-open, in an attempt to hit the eyes of its enemy. This capacity is absent in other subspecies of the Indian Cobra.

This cobra is preyed upon by mongooses, which eat its eggs as well, and by large monitors and raptors. However, neither the mongooses nor the other animals are totally immune to the poison, although some are highly resistant.

BANDED KRAIT
Bungarus fasciatus

The Banded Krait ranges from India to southern China and south-eastern Asia. It can attain a length of 175 cm, but averages 150 cm. It is nocturnal, sheltering by day in crevices and holes, and coming out after nightfall to hunt other snakes, lizards, frogs, and occa-

Banded Krait

Russell's Viper

Halys Pit-viper

sionally fishes and small rodents. The Banded Krait is very shy and non-aggressive during the day. Small children have been known to catch a krait and play with it, without the snake attempting to use its dangerous venom fangs. If it is disturbed during the day, it pushes its head underneath its body, or into a hole or among the grass to protect itself from the attacker. It can form circular coils with its body, jerking them violently and creating the effect of a large colourful disc. This often deters the assailant. Sometimes people are bitten by a krait when they have stepped on it barefoot. Such accidents often prove fatal. The female lays 10—15 eggs in holes, and the young hatch after about 2 months.

RUSSELL'S VIPER
Vipera russelli

Russell's Viper is a large snake, up to 1.75 m long, with a stout body. It is distributed in India, Sri Lanka, Burma, Thailand, Sumatra and Java. It occurs in three subspecies differing in coloration, and is very common in some localities. It is often found in gardens and has been caught in the parks of Calcutta, though it usually frequents areas of scrubland. When it is disturbed it never retreats but inflates its body and makes angry hissing sounds. Russell's Viper does not attack man, but its venom is very strong, and when it bites a person who inadvertently steps on it, the bite will probably prove fatal unless an antidote is administered. Russell's Viper is nocturnal, taking refuge always in the same holes by day. It feeds mainly on small rodents. The female bears about 10 live young.

HALYS PIT-VIPER
Agkistrodon halys

The Halys Pit-viper has an extensive area of distribution stretching from the river Volga across central Asia to Siberia and the Far East, and also covering Mongolia, China, Japan and Thailand. It reaches a length of about 60 cm. Several subspecies, differing in coloration and pattern, are distinguished within its territory. Some have small dark spots on their flanks. The Halys Pit-viper is usually found in grassland and bush-covered areas, but it can also be encountered up to a height of 4 000 m in the mountains. It lives among stones or beneath bushes and is abundant in some localities. It rests in a secluded place by day, coming out to hunt in the evening. It preys on small rodents, small ground-nesting birds, and lizards. Its venom is quite strong and can be dangerous to children or less vigorous adults. The female bears 6—8 live young in summer.

SCORPION
Thelyphonotus caudatus

The scorpion *Thelyphonotus caudatus* is confined to stony and sandy localities in southern Asia. This scorpion reaches a length of 3.5 cm and has a long appen-

Scorpion Thelyphonotus caudatus

Scorpion Leiurus dufoureis

dage on its abdomen. It makes its home beneath stones or in burrows dug in the sand, and here it drags its prey, consisting mostly of tiny insects. The female lays 20—35 eggs in a special pouch carried under her abdomen. The young remain on their mother's back for a week, by which time they have undergone their first moult.

SCORPION
Leiurus dufoureis

The scorpion *Leiurus dufoureis*, which measures up to 7 cm, inhabits stony tracts of desert and semidesert in Turkey. During the day it shelters beneath large stones seeking the small amount of moisture which is vital for their survival. It begins to hunt after nightfall, slowly probing its territory in search of insects and spiders. Small prey is squeezed to death in the pincer-like claws, but if the prey is rather larger, the scorpion uses the venomous sting situated at the tip of its tail. It then proceeds to chew the prey, decomposing the flesh by means of its digestive juices, and sucking in the resulting liquid. The female bears about 30 live young. They immediately climb onto their mother's back and stay there for a week until their first moult has taken place. After this the young scurry about over their mother's body, finally leaving her after a few weeks have passed and they have become independent. Until this time the mother continues to feed them. The sting of this scorpion is not particularly dangerous for man, but the wound may be swollen and painful.

SPIDER
Galeodes orientalis

The spider *Galeodes orientalis* inhabits stony and sandy areas in the Asiatic tropics and subtropics. It reaches a length of 5 cm and has a hairy body and huge pincer-like mandibles which move vertically. They are extremely powerful and the spider can easily kill a small lizard, though it is not dangerous to man, being a non-venomous species. However, the local people fear it more than they fear poisonous snakes. The diet of this spider consists mainly of insects, especially locusts and bugs found on the ground, termites and even molluscs. This species shelters underneath large stones during the day and comes out to hunt at dusk.

Spider Galeodes orientalis

Greenhouse Camel Cricket

Moth Phyllodes verhuelli

GREENHOUSE CAMEL CRICKET
Tachycines asynamorus

The Greenhouse Camel Cricket, which is about 18 mm long, is found in the warm regions of eastern Asia. It has hair-like antennae greater in length than its body, and the female has a long ovipositor. During the day the cricket remains hidden underneath stones or in a crevice, coming out after nightfall to prey on other insects and nibble at shoots. The female lays about 400 eggs in the ground, and the newly hatched larvae undergo several moults before they are fully grown. This cricket was accidentally introduced to Europe, where it settled, and multiplied, in warm greenhouses.

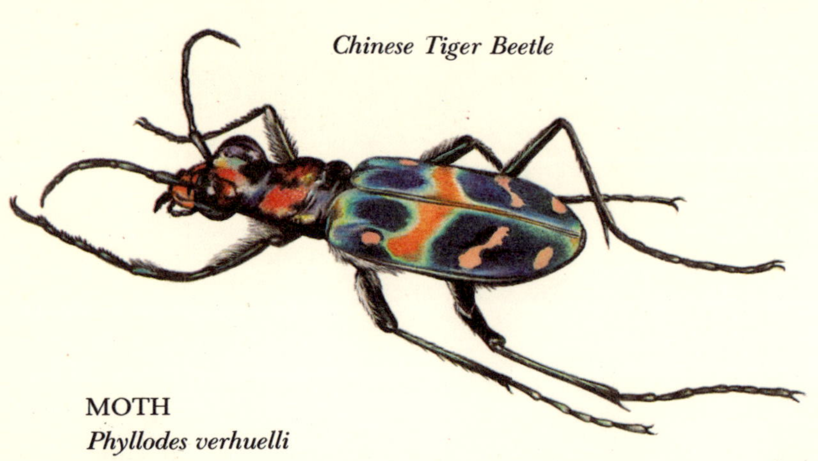

Chinese Tiger Beetle

Desert beetle Anthia bucharica

MOTH
Phyllodes verhuelli

Phyllodes verhuelli, one of the world's largest moths, is confined to eastern India and the Andaman Islands. It has a wingspan of up to 15 cm. This huge moth rests on the trunks of trees or bushes where its coloration blends with its background. When disturbed, it spreads its wings to reveal large eye-like spots on the hind wings. These frequently alarm potential predators such as birds or lizards.

CHINESE TIGER BEETLE
Cicindela sinensis

The Chinese Tiger Beetle, which reaches a length of about 20 mm, is one of the most beautiful species of beetle. It is indigenous to China, where it lives in sandy or stone-strewn localities. It scurries speedily over the ground, catching small insects and spiders, and when it is disturbed it flies up immediately only to land again a few metres away. On rainy or overcast days, it takes refuge in holes or underneath stones. The female lays her eggs in the ground. The predatory larvae dig vertical tunnels, sometimes over 1 m

deep, in the soft soil. Then they lie in wait at the top and prey on insects passing by. They catch their prey with their strong mandibles and drag it inside the tunnel where they devour it. Larval development usually takes 2 years.

DESERT BEETLE
Anthia bucharica

The desert beetle *Anthia bucharica* is common in some parts of central Asia. This beautiful predatory beetle, which measures up to 60 mm, has extremely powerful mandibles. It frequents sandy localities covered by scrub and thin grass, usually near a source of water or a swamp. During the day it hides away in holes dug out in the sand beneath the bushes. It comes out to hunt in the early morning and in the evening, running on the sand with speed and dexterity. It preys on insects, worms and molluscs, eating only the soft parts of them. It chews the flesh, and uses a special secretion containing butyric acid to liquify it so that it can suck it in. The female lays her eggs in the ground, the hatched larvae being just as predatory as the adults.

WATERS
AND
SWAMPS
OF THE WARM
AREAS

———

The huge, permanent glaciers of the soaring Himalayas, the Tibetan plateau and other Asiatic massifs for most of the year feed the largest rivers of southern Asia. In June, when the glaciers and the snow begin to melt, the rivers rise sharply, and in June and July the monsoon rains cause considerable flooding over vast low-lying areas. Many rivers, often by this time several kilometres wide and driven by powerful, uncontrollable masses of water, change their courses as they flow. The Ganges once flowed directly southwards, following the courses of the smaller present-day rivers Bhagirathi and Hooghly, and Brahmins still regard the western branch as the true, sacred Ganges. In the 16th century, the Ganges began to push south-east, and its waters merged with those of the Brahmaputra, to make one massive river, which, even now, does not have a fixed bed. In the vast delta which extends far inland, the branches change their courses almost every year. When the rate of flow is at its greatest, the combined rivers voraciously grind out the soft river bed, creating new beds and carrying sediment with them, to be deposited further down the river, especially in the river delta. The sediment forms natural dams, which are soon reinforced by a variety of plants, and the streams are compelled to take new courses and different directions. The Ganges contains an enormous quantity of alluvium. Each cubic metre contains about 200 g of sediment when the flow is normal, and up to 2 500 g during the rainy season. The water is always brown and muddy, being coloured by the soil torn from the banks. In contrast, the streams at the source of the river are crystal clear as they gush from the Himalayan glaciers more than 3 000 m above sea level, and they are as sacred to the local people as the river itself. After a distance of 2 700 km, the Ganges flows through a densely branched delta into the Bay of Bengal. Its huge tributary, the Brahmaputra, also rises in the Himalayas. It forges wide valleys along its entire 2 960 km length, and these are often inundated after heavy rains.

Another river of southern Asia, the Sind, formerly known as the Indus, contains an even larger percentage of sediment. Its source is in the Himalayas, on the Tibetan border, and it enters the Arabian Sea by way of a wide delta south of Karachi. Its usual content of alluvium is 2 500 g per cubic metre of water, but when the river is in spate, it increases to 4 500 g. The

Sind often changes its course and deposits thick sediment on its flat banks. When the mountain snow begins to melt, a vast area along the 3 190 km length of the river becomes submerged. The stream noisily tears enormous chunks from the banks, and the racing water carves out a new river bed, often several kilometres from the original one. Soon many villages and townships are inundated, and this often results in heavy casualties when the inhabitants are not evacuated. There is no way in which this destructive element can be contained, although flooding happens almost regularly, in July. The rivers are at their highest during the season of monsoon rains, and the water level falls in September and October. Very little water flows in winter, and some smaller tributaries dry up almost completely, leaving only swamps. For most of the year the basin of the Sind is surprisingly arid, and soil only 50 m or so from the stream is parched. In some areas, date palms which are normally typical of a dry climate, grow along the banks of this river.

In some regions, the monsoon rains often come unexpectedly in November, causing floods which ruin the crops.

The Mekong River in south-east Asia is another huge waterway. Its source is in Tibet, at an altitude of over 3 000 m, where it is fed from glaciers and melted snow in spring, and by the monsoon rains in summer. The upper part of the Mekong River flows through deep ravines, over foamy rapids. Further south, the river widens as it is joined by its tributaries, and from the Chinese border it is navigable by large boats and barges. It flows through Laos and Cambodia, and in southern Vietnam, after a journey of 4 500 km, the second longest river in Asia flows into the South China Sea.

The Yangtze River in China with a length of 4 980 km is the longest river in Asia, and it is second only to the Ganges in its rate of flow. Another Chinese river, the Huang-Ho, with a length of 3 343 km is also one of the longest Asian rivers. Both rivers rise in Tibet, receiving water from glaciers and from monsoon rains, and they both run east to the Yellow Sea. Other large monsoon rivers of south-east Asia are the Red River, the Black River, the Irrawaddy, and the Salween.

These tropical rivers have a considerable influence upon the life of the native populations of their vast basins. There are many channels both man-made and natural between the main rivers and their side branches. These are used for transportation of merchandise, for fishing, and to irrigate the field during the dry season. There are all kinds of boats, barges, canoes and rafts on these waterways. In deeper water large sailing boats carry timber, bamboo, corn and cattle from distant regions. Houseboats roofed with bamboo and palm leaves, and anchored in the backwaters, provide shelter for many families. Small fishing boats propelled by oars traverse both the large rivers and the smaller branches. Fish is one of the principal components of the local people's diet. Fishermen sell colourful fishes of all shapes and sizes in the markets, and it is also possible to buy freshwater turtles, mostly of the genus *Trionyx,* which have become trapped in the fishing nets. The fish merchants usually cut off the turtles' heads, for a large turtle can easily snap off a customer's finger with its sharp jaws. Freshwater crabs are also sold in the markets. In south-eastern Asia, they are tied by the dozen, with bamboo fibres, to improvized ladders, also made of bamboo. This is because the crabs bite off each others limbs — including the claws — when they are put together in a basket. Fishing is also frequently carried on in mountain rivers and in lakes. Throughout the centuries, local people have used traditional canoes made from huge tree trunks.

Large numbers of animals obtain their food from tropical waters. In winter, the rivers, lakes and swamps become the home of water and swamp birds migrating from more northerly parts of Asia. Herons, marabous and ibises build their colonies in trees along the edges of the rivers, while various species of ducks settle in the thick vegetation beneath the trees. Wading birds use their long legs to make their way through the shallows where they catch fishes and aquatic molluscs, such as *Melania tuberculata* and *Corbicula orientalis* or small racing crabs of the genus *Ocypode.* The river deltas are favourite haunts of crocodiles, and in the River Ganges the 'sacred' Indian Gavial is to be found.

Herds of wild water buffaloes are frequently encountered on the swampy ground near rivers, and the rare and strictly protected Indian One-horned Rhinoceros is found only in the Indian

swamps. During the last century this species had a wide area of distribution, but it is now found in only a few localities, such as the Kaziranga National Park in Assam, where there are 400 specimens. Another 300 rhinoceroses live outside this reserve. Small herds of elephants often visit pools in swampy areas, or rivers and lakes in places remote from civilization.

The landscape around rivers and lakes has its own characteristic trees and bushes. Typical of the Indian region are palm trees of the genus *Borassus,* the fruits of which are up to 15 cm long and contain three stones in the pulp underneath the thin shell. There are many tall, deciduous trees of the genus *Terminalia.* The Indian species *Terminalia chebula* yields edible oily seeds, while the Indian almond *Terminalia catappa* of Malaysia is the source of oil obtained by pressing the almond-like fruit. India also has several species of the genus *Albizzia.* The East Indian Walnut *(Albizzia lebbeck)* has large crowns and purplish-pink blossoms, and it yields high quality dark brown timber. The Tamarind *(Tamarindus indica)* is a large tree which grows near water in the tropics. The pulp inside its thick pods is regarded as a valuable source of food. Rosewoods of the genus *Dalbergia,* characteristic of hot, wet areas of south and south-east Asia, also have pods, and *Dalbergia latifolia* yields pink, or sometimes dark purple to blackish timber of high quality. Trees of the genus *Mangifera* flourish in irrigated or regularly flooded regions. Especially popular is the Mango *(Mangifera indica),* which has tiny yellow blossoms and ovoid fruit, about 10 cm long, which weigh up to 1 kg. The greenish or yellow fruit is highly nourishing and has a pleasant-flavoured, sugary flesh, from which the juice is obtained.

Quiet backwaters, lakes and swamps are in many places rimmed with extensive, dense beds of tall reeds, particularly the species *Phragmites karna.* Here a variety of birds and even large mammals are able to find shelter. In central Asia, reed beds cover the banks of the irrigation canals. Reeds spread easily in swampy ground and the tips of the roots can pierce the harder soil several metres from the bank. Reeds also form floating islands when patches torn away by torrential rains are carried by the current to be cast up in quiet river coves. In southern Asia reeds are used for many purposes. They cover the walls of huts, and they are woven to make rugs, fences, and so on. Rushes of the genus *Cyperus* grow to a height of 3 m in the swamps, and small weaver birds nest among the dense growth. The plant *Colocasia antiquorum,* which reaches a height of 150 cm, has tufts of long leaves, and a white, funnel-shaped sheath or spathe surrounding a stout yellowish spike or spadix, the lower part of which conceals tiny male and female flowers. Its tubers, which weigh up to 2 kg, yield a starchy material called taro. They also contain toxic substances, but these are eliminated during cooking. The tubers are an important source of food in India, from where the plant has been introduced into tropical areas throughout the Old World. The Sweet Flag *(Acorus calamus),* another plant of Indian origin, grows in the mud beside the rivers. Its rootstock yields an aromatic drug, and this plant has similarly been distributed throughout the northern hemisphere.

Many other plants of the humid tropics produce colourful flowers and are often grown in greenhouses in cooler parts of the world. They include plants of the genus *Typhonium,* which have tufts of heart-shaped leaves and purplish, funnel-shaped spathes. When they are in flower, their intense smell is reminiscent of decomposing carrion and it attracts many species of flies which aid in pollination. The Indian plant *Amorphophallus campanulatus* has a visually impressive, bell-shaped spathe, up to 30 cm long, green and white outside and dark violet within. The spathe of the related *Amorphophallus titanum* of Sumatra reaches the incredible height of 2 m.

Many shallow reservoirs and areas of slow-moving water are covered with water hyacinths of the genus *Eichhornia.* This water plant has magnificent pale violet flowers, and for this reason it was introduced as an ornamental plant in southern Asia in the early 20th century. It acclimatized successfully, but it reproduced so rapidly that many stretches of water became completely choked by it.

Turmeric *(Curcuma longa)* is cultivated in moist areas of Indo-China. Its rich yellow rhizomes are powdered to give the condiment turmeric, which is one of the principal components of curry powder.

Of all the plants which grow in swamps Rice *(Oryza sativa)* is the best known. This important crop is native to southern and eastern Asia, where it still grows in the wild. It is an annual grass, 50—120 cm tall, with a head of flowers, and unlike most other grasses, it requires contin-

uous irrigation. It is cultivated in irrigated paddy fields, or on terraced hillsides up to a height of 2 000 m. In the Himalayas and other mountains of the tropics, water is carried to the fields in pails. In lowland areas or in mountain valleys, water is often pumped with the aid of domesticated buffaloes. Sometimes a system of large wheels is constructed, in which paddles scoop up water from a stream and pour it into troughs from which it flows to the fields. Rice ripens after 6—10 months, and in warm regions it can be harvested twice a year. The fine rice straw is used to make brooms, brushes, rugs and hats, and in China it is used in the manufacture of paper. Rice was grown in Asia in prehistoric times, and over the centuries its cultivation has spread to tropical and subtropical areas throughout the world. At the present time several hundred varieties are grown. Rice paddies provide habitats for a variety of aquatic invertebrates, and for the animals which feed on them. Here herons find easy pickings and do no damage, but at harvest time, large flocks of seed-eating birds arrive and wreak havoc in the rice paddies. Most damage is caused by the Java Sparrow *(Padda oryzivora)*. These sparrows are indigenous to Java and Sumatra and were once frequently kept as cagebirds. Many escaped into the wild and so spread to other parts of the world. The species is now found throughout south-eastern Asia, Japan, east Africa and Hawaii, the birds taking up residence in villages and nesting in nearby bushes. As soon as the rice crop ripens, enormous flocks form, which descend on the paddy fields and ruin the harvest.

In all the tropical regions of Asia which are rich in water, man has to face one of the greatest hazards to health—malaria. This disease is carried by some species of mosquitoes, which flourish in moist, warm sites, where they can reproduce all year round. After sunset, swarms of mosquitoes emerge, and in some places beds have to be covered with fine mosquito nests to keep the insects out.

RUSSIAN DESMAN
Desmana moschata

The Russian Desman grows to a length of about 42 cm, of which 20 cm are taken up by the tail. The tail is flattened from side to side and is thickened at the base. It is ringed with scales and has a covering of sparse hair. The desman lives in the river basins of the Don, Volga, Dnieper and Ob, and has been introduced in other places for its valuable fur. It digs burrows in the river banks and builds underwater entrances. As many as 8 adult desmans live together in one underground chamber. If the water-level falls too far in the dry season, the desman moves to another river. It moves speedily and nimbly in the water, using its tail as a paddle or a propeller, but it can also use its legs to swim. Desmans hunt after nightfall preying on aquatic animals, particularly fishes, but also on crustaceans, molluscs, amphibians and insects. After a gestation period of 50—60 days, the female bears 1—5 young in her burrow. Young desmans are born blind and naked, opening their eyes after 3 weeks. They are suckled for 1 month and become independent at 6 weeks, although they are not fully grown until they are a year old. The Russian Desman was at one time much hunted for its fine and durable pelt, but it now enjoys strict protection in most of the areas of its distribution.

ORIENTAL SMALL-CLAWED OTTER
Amblonyx cinerea

The Oriental Small-clawed Otter is widespread from India to southern China and south-eastern Asia, extending to the Greater Sunda Islands and Palawan. It reaches a length of 90 cm, a third of which is taken up by the tail. An adult otter weighs 2.7—5.4 kg. This species frequents rivers and lakes in hilly areas, but it also abounds along the coasts and on offshore islands. During daylight and at dusk it hunts in the water for fishes and crustaceans, and catches small birds in the reed beds. Sometimes it finds birds' eggs on the banks. In places where it is hunted during the day, it becomes active at night. Following a gestation period of 60—63 days, the female gives birth to 2—4 naked and blind young. These open their eyes after 28—35 days and take their first solid food at 8 weeks of age. They are weaned when they are 4 months old by which time they can dive and swim skilfully, and are learning to hunt. The Oriental Small-clawed Otter is prized for its fine fur.

MALAYAN TAPIR
Tapirus indicus

The Malayan Tapir is indigenous to Sumatra, Malaysia and Thailand. It is a robust animal, standing 130 cm high at the shoulder, measuring 250 cm in length and weighing up to 300 kg. It lives in primary forests in river basins, or in swamps where there is lush vegetation, ranging from low-lying land to mountainous areas at a height of 4 500 m. The tapir pushes its way speedily and skilfully through thickets and reed beds. It is a good swimmer and can dive when it is in danger. By this means it can escape its principal enemy, the tiger. Tapirs are solitary for most of the year, only briefly forming pairs. Breeding can take place at any time of the year. After a gestation period of 390—405 days, the female bears a single young in a dry, sheltered place in the undergrowth. The young tapir weighs 10 kg at birth. It has a dark ground colour with white and yellow stripes, serving as protective coloration in the bamboo or reed jungles. Females mature after 3—4 years, males after

Oriental Small-clawed Otter

Malayan Tapir

5 years. The Malayan Tapir feeds on leaves, twigs, grass, fruit collected on the ground, and on aquatic plants collected beneath the water. It browses during the day or at dusk, but in places where it is often disturbed or hunted, it feeds only at night.

INDIAN ONE-HORNED RHINOCEROS
Rhinoceros unicornis

The Indian One-horned Rhinoceros lives at present in the protected territories of the Kaziranga National Park in Assam, in the Jaldapare National Park in Bengal, and in a few areas in Nepal. This very rare species once ranged from Kashmir to Assam and northern India, but over the centuries it was ruthlessly hunted for its horn and its blood. The horn, which consists of hair-like fibres cemented together, was ground to a powder, and together with dried blood was exported to China, where is was paid for in gold. The Chinese believed that both had magical powers, and used them widely in their medicines. As a result rhinoceroses became progressively more rare until governments were compelled to take protective measures. Only a few hundred of these animals now survive in the wild. The Indian One-horned Rhinoceros frequents swampy jungles containing bamboo thickets and elephant grass, especially in localities where there are small lakes or rivers where it can swim, for it is an accomplished swimmer and diver. The rhinoceros normally lives in pairs or in groups made up of families with their offspring, but old males are sometimes solitary. On their regular daily forays, the animals tread out paths, or even form tunnels in the tall, dense vegetation. They occasionally feed at dusk, their diet consisting of young grass, reeds, bamboo leaves

Indian One-horned Rhinoceros

Javan Rhinoceros

and shoots, twigs and aquatic plants, particularly water hyacinths.

After a period of 462—489 days, the female bears a single young, or occasionally twins. The young rhinoceros weighs about 60 kg at birth and can stand up after 10—60 minutes. It starts to suckle after 3 hours and is weaned after 1.5—2 years, although it can nibble grass when it is 2 weeks old. It takes its first drink of water at 3.5 months. An adult rhinoceros stands up to 200 cm high at the shoulder, is 3—4.5 m long and weighs 2 000—4 000 kg.

JAVAN RHINOCEROS
Rhinoceros sondaicus

The Javan Rhinoceros was once distributed from East Bengal across south-eastern Asia to Sumatra and Java, but now only 40 animals remain, living in the Udjong-Kulou Reserve in western Java. The Javan Rhinoceros is about 150 cm high at the shoulder. The male has a single small horn, which is absent in the female. This species inhabits tropical forests and jungles near lakes and swamps. Rhinoceroses also live along the banks of rivers and often bathe, remaining

in the water for several hours, particularly in the heat of the day. They feed in the morning or afternoon, browsing on leaves and twigs of bushes or low trees. Except in the mating season, rhinoceroses are solitary. The male roams through the countryside, but the female with her single young remains in her own territory. She gives birth after a gestation period lasting 470 days. The young is suckled for 1.5 years, and drinks as much as 15 l of milk a day for the first few months.

SUMATRAN RHINOCEROS
Dicerorhinus sumatrensis

The Sumatran Rhinoceros is the smallest species of rhinoceros, reaching a height of 130 cm at the shoulder, a length of 280 cm, and a weight of 1 000 kg. Its two horns distinguish it from the other Asiatic species. In males the front horn is 45 cm long, and the back horn measures 15 cm. In females, the horns measure 15 cm and 5 cm. The Sumatran Rhinoceros once occurred in great numbers throughout south-eastern Asia, Sumatra and Borneo, but it has been almost exterminated as a result of ruthless hunting. It is pres-

Sumatran Rhinoceros

196

Chinese Water Deer

ently confined to a few localities in southern Sumatra, northern Borneo, Thailand and Burma, but altogether only about 100 of these animals still survive. The Sumatran Rhinoceros frequents mountain jungles near water. At night it sleeps in thickets, preferably in the middle of a swamp, and it comes out to feed after sunrise. At midday it rests in the shade of trees or bamboo thickets, browsing through the afternoon on twigs and leaves of bushes, trees and bamboos. It is usually solitary outside the mating season. After a gestation period of 17—18 months, the female bears a single young, which stays with her for more than 2 years.

CHINESE WATER DEER
Hydropotes inermis

The Chinese Water Deer is found in the drainage basin of the lower Yangtze and in Korea. It is 45—55 cm tall at the shoulder, about 100 cm long, and it weighs 9—11 kg. Antlers are absent in both sexes, but the male has elongated upper canine teeth, forming curved tusks which it uses in self-defence and with which it can inflict deep wounds. The males engage in fierce fights during the rutting season, in December or January. After a gestation period of 180—200 days, the female bears 2—7 white-spotted young in a bush-covered shelter. The young remain with their mother in the seclusion of their territories. The Chinese Water Deer is usually found among tall reeds and grasses, and along the edges of forests, usually near water. It also frequents cultivated fields and plantations. It forages during the day, feeding on swamp and water plants.

WATER BUFFALO
Bubalus arnee

The Water Buffalo is one of the best-known animals of the Asiatic tropics and subtropics. Its area of distribution ranges from India and Sri Lanka to Nepal, Assam, south-eastern Asia and Borneo. It is found in the wild in only a few areas, but the domestic form is bred from Egypt to the Philippines. The Water Buffalo reaches a height of 180 cm at the shoulder, a length of

Water Buffalo

197

300 cm, and a weight of 720—820 kg. Its tail measures 100 cm in length. The bull has robust horns, up to 120 cm long, while the more slenderly-built cow has more delicate horns. This buffalo inhabits large swamps and the edges of rivers and pools, being found up to a height of 1 800 m in the mountains. It is gregarious, forming herds which defend their own territory. Each herd is made up of 10—12 cows, and in the rutting season the bulls fight for possession of these herds. When the rutting season is over, the males either become solitary or become members of large herds. After a gestation period of 287—340 days, the female bears a single young, able to stand after 30 minutes and to suckle within an hour. It is suckled for 6 months, but can graze from the time it is four weeks old. Water buffaloes feed on leaves, twigs, grass, the shoots and leaves of bamboo, and on swamp and water vegetation. They feed mainly in the early morning and late afternoon, wallowing in mud or in shallow water during the heat of the day. Herds made up of cows with their calves are not afraid to attack even so fearsome a creature as a tiger, if it should chance to threaten them. Indeed, this largest of the Asiatic carnivores always flees before a herd of enraged buffaloes.

Great Cormorant

GREAT CORMORANT
Phalacrocorax carbo

The Great Cormorant, which reaches a length of 90 cm, inhabits large lakes, lagoons and rivers throughout the temperate and tropical zones of Asia. Outside the nesting season, it wanders in small groups of 8—10, but it nests in large colonies, those of northern India often comprising several thousand pairs. The nesting season in this area lasts from September to December. Nests are built by both partners from twigs and sticks, broken off with the birds' powerful beaks, and they are constantly improved, mainly by the male, even after the eggs have hatched out. There are sometimes scores of nests in a single tree with the result that trees become defoliated within a few years, adding to the destruction caused by excrement which damages the bark. It is not unusual to see river banks and the shores of lakes littered with dead barkless trees. In some localities, notably on the banks of the Subansiri in Assam, cormorants build their nests on rocks. The female lays 3—6 pale bluish-green eggs and sits on them alternately with the male for 24 days. The newly hatched young are blind and naked. They open their eyes after 3 days, and are soon covered with a layer of black down. They peck their food from the gullets of their parents, who bring them tiny fishes, crustaceans and tadpoles. Cormorants catch their prey underwater, hunting in groups, and often rounding up the fishes. At the age of 5 weeks, young cormorants leave the nest to perch on nearby branches. They take their first flights when they are 2 months old.

GREY PELICAN
Pelecanus philippensis

The Grey Pelican reaches a length of 150 cm. It is widespread on all large watercourses and lakes in south and south-east Asia, in the Philippines, Sumatra and Java. It also lives in coastal regions, particularly outside the nesting season, at which time pelicans from some regions travel long distances. They are good fliers, and are capable of sustained flight, their V-shaped formations often being seen high in the sky on sunny days, though pelicans also fly in straight line formations.
The large nest, measuring about 1 m across, is made of branches, reed stems and grass. It is situated in the

branches of a large tree growing in a marshy area.
There are usually several pelican nests in one tree,
along with nests of herons and storks. Pelicans often
nest in tall palms trees, building their nests at the
base of the fan-shaped leaves. The female lays 2 whitish eggs, which are incubated by both partners for one
month. Both parents feed the young, at first regurgi-
tating tiny fishes into the nest. Later the young pick
food directly from the adults' pouches. Young pelicans
begin to fly after 5 months, but they often swim with
their parents when they are 2 months old.

BLACK BITTERN
Dupetor flavicollis

The Black Bittern is found throughout south and
south-east Asia, the Greater Sunda Islands, the Philippines, and Australia. This wading bird is about 55 cm
long. It frequents rivers, swamps, rice paddies and
coastal mangrove swamps, and is found on mountain
lakes up to a height of 1 200 m. During the day, it
hides in thick vegetation, coming out after nightfall to
catch fishes, frogs, molluscs, crustaceans and insects.
After the young have fledged, the Black Bittern lives
a solitary life, but it forms pairs in the nesting season,
at which time the male makes a loud, lowing call.
Both partners participate in building the nest, which
is made of twigs and water plants. It is well-concealed
in a reed bed in the marshes or near the ground
among bamboo stalks. The female usually lays 4 white
eggs with a bluish or greenish sheen. She begins incubation as soon as the first egg is laid and the male relieves her regularly. The young hatch after about
18 days and stay in the nest for 12 days or so before
scattering in the neighbourhood and climbing among
the reeds, bushes and bamboo thickets. The adults
bring food, which they at first regurgitate into the
nest. Later they feed the prey directly to the young.

CHINESE POND HERON
Ardeola bacchus

The Chinese Pond Heron is distributed from eastern
Assam to China, and across south-eastern Asia to
Borneo. It is about 52 cm long. It also visits Japan,
but does not breed there. This heron lives in flocks
among swamps, on lake shores and along the coasts,
and it is common in rice paddies. It catches fishes,

Grey Pelican

Chinese Pond Heron

Black Bittern

Japanese Crested Ibis

JAPANESE CRESTED IBIS
Nipponia nippon

The Japanese Crested Ibis is about 75 cm long. It is one of the most beautiful and also one of the most rare of the wading birds. It is native to the Japanese islands, Korea and eastern China, where it lives in small groups on the shores of lakes and rivers in deciduous woods. Each pair builds a nest of twigs, grass stalks and other plant materials, in a treetop. The female lays 2—4 bluish-green eggs and takes turn with her partner in sitting on them for 22 days. The young are fed by both parents. The diet of the Japanese Crested Ibis is made up of insects, molluscs, crustaceans and frogs.

GLOSSY IBIS
Plegadis falcinellus

frogs, crustaceans and aquatic insects in shallow water, and on the coastal mudflats it usually preys on mudskippers and ghost crabs. On grassy sites near water it also catches locusts and grasshoppers. Herons often perch on the backs of buffaloes. Here, they not only have a good view of their surroundings, but they are able to prey on the many small creatures disturbed by these massive animals as they walk. In return the herons pick insects and insect larvae from the buffaloes' skins. The Chinese Pond Heron nests in small groups made up of just a few pairs, or in colonies together with other species of heron. The nest is built of twigs, sticks and pieces of reed. Both partners incubate the clutch of 3—5 pale green eggs for 24 days, and later both feed the young.

The Glossy Ibis is a cosmopolitan bird occurring in every continent. In Asia its area of distribution stretches from Pakistan across the whole of south and south-east Asia to the Philippines and the Sunda Islands. It is migratory in the more northerly regions, but otherwise it is resident. It lives beside lakes and in swamps, in coastal lagoons and in mangrove jungles. This wading bird, which is about 65 cm long, moves in groups and nests in colonies, often with herons. The nest of sticks and grass stalks is built in bushes or trees in the middle of the swamps, and the clutch of 3 dark greenish-blue eggs is incubated by both birds for 21 days. The young are fed by both parents. The Glossy Ibis feeds on worms, molluscs, crustaceans, insects and frogs.

Glossy Ibis

PAINTED STORK
Ibis leucocephalus

The Painted Stork, which is about 100 cm long, is found in India, Burma, Thailand, southern Vietnam and in the south and east of China. It frequents lakes and swamps, but it is also common in rice paddies, where it catches fishes, crustaceans, frogs and aquatic insects in shallow water. In the evening, it roosts in groups in tall treetops, and it nests in large colonies, often with other wading birds. The nest is built in a tall tree standing in water, or among the mangroves. Acacias and mesquites are popular nesting trees, as are *Barringtonia racemosa*, and solitary fig trees, such as the Bo Tree *(Ficus religiosa)*. There are usually more than 20 nests in the same tree. The nest is made of twigs, which the birds pick out of water with their beaks, and it is lined with various water plants. The female lays 2—5 white eggs, occasionally with brown spots. The clutch is incubated by both birds for 28 days, and both parents feed their young.

ASIAN OPEN BILL
Anastomus oscitans

The Asian Open Bill is a species of stork, about 80 cm long, characterized by its unusual beak, in which there is a gap between the mandibles. It is native to India, Sri Lanka, Burma, Cambodia and southern Vietnam, being abundant in some localities. Outside the nesting season it travels in pairs or singly, later gathering in huge colonies of up to several thousand birds to breed. The nest is built in a tall tree, especially in acacias, mesquites and trees of the genus *Barringtonia* in the middle of extensive swamps or beside lakes or rivers. There can be more than 30 large nests made of twigs and lined with leaves, in the branches of a single tree. When the partners meet in the nest, they greet each other by clattering their mandibles, and sometimes they make deep lowing sounds. The female lays 2—4 whitish eggs, which the parents incubate for 24—25 days. They feed their young on freshwater gastropods, particularly on large specimens of the species *Pila globosa*, the birds skilfully cracking the shells with their bills. They can also open clams, pecking out the bodies with the tip of the bill. Open bills also catch crabs, frogs and other small creatures of the swamps. The nestlings have normal bills, but as soon as they begin to hunt for themselves, the mandibles open and the characteristic gap appears.

BLACK-NECKED STORK
Xenorhynchus asiaticus

The Black-necked Stork is indigenous to India, south-eastern Asia, New Guinea and Australia. This long-legged bird attains a height in excess of 170 cm and a length of 135 cm. Both sexes are alike in colour, but the male has a brown iris, while that of the female is lemon yellow.
This stork inhabits swamps, rivers and lakes, where it

Painted Stork

lives for most of the year in small groups made up of families with their offspring. In the breeding season it forms pairs and defends its territory, the nesting season varying with the area of distribution. The nest is built by both partners in a treetop, often in the Bo Tree *(Ficus religiosa)*, 20—25 m above the ground, or in a tall banana tree. The female lays 3—5 white eggs

Asian Open Bill

Black-necked Stork

and sits on them alternately with the male for 33 days. Both parents feed their young. The Black-necked Stork feeds on fishes, frogs, reptiles, crabs and other small animals.

LESSER ADJUTANT STORK
Leptoptilos javanicus

The Lesser Adjutant Stork inhabits south and south-east Asia and the Greater Sunda Islands. It reaches a length of 120 cm. Outside the nesting season it leads a solitary life, wandering in swampland, rice paddies, on the edges of lakes and rivers, or in mangrove swamps along the coasts. There it hunts fishes, frogs and crustaceans, killing them on land with its strong bill. It also catches lizards, snakes and large insects. Several of these storks sometimes gather at the carcass of a large mammal, pecking at the flesh, entrails and skin.

During the breeding season partners often hop from one foot to the other, as if they were tap-dancing, as part of their courtship display. They build a huge nest up to 200 cm across and 100 cm tall, in woodland at the water's edge, usually in silk-cotton trees, 12—30 m above the ground. The nest is made of dry sticks and branches, and is lined with pieces of hide, hair or grass. The female lays 3—4 white eggs, which are incubated by both parents. They both feed the young, which hatch out after 35 days.

RUDDY SHELDUCK
Tadorna ferruginea

Lesser Adjutant Stork

The Ruddy Shelduck is about 66 cm long. It lives in an area around the Black and Caspian Seas, extending east to China. Outside the breeding season it flies to the rivers and lakes of south and south-east Asia. It is less gregarious than related species, living either in pairs or in small flocks of 20—30. However, as many as 15 000 shelducks have been observed on Lake Chilka in the migrating period. Shelducks return to their nesting grounds in pairs, and settle near lakes up to a height of 4 000 m, or even at 5 000 m in Nepal. The nest is built in a burrow vacated by foxes, or in heaps of stones or in fallen hollow trees. The female lines the nesting cup with leaves and pieces of wood, and surrounds the clutch with down. She lays 6—10 ivory white eggs and sits on them for 28—30 days, while the male stays on guard nearby. As soon as the newly hatched ducklings have dried, their mother takes them to the water, where they feed independently. The Ruddy Shelduck eats seeds, green plants, worms, molluscs, aquatic insects and occasional tadpoles.

SPOTBILL DUCK
Anas poecilorhyncha

The Spotbill Duck attains a length of 60 cm. It is widespread from Pakistan and India to north-east, east and south-east Asia and the Philippines, migrating outside the nesting season to Sri Lanka, Cambodia and Thailand. In Pakistan, this bird of rivers, lakes and swamps lives mainly on the Sind, and in Kashmir it occurs on mountain lakes at a height of 1 800 m. When the nesting season is over, it travels both in small flocks and in large groups of over a hundred.

It forms pairs before arrival at its nesting grounds. Courting partners swim around each other, the male turning his bill down, ruffling his feather and jerking his tail as he moves his head to and fro, and dips his bill into the water. The female is similar in colour to the drake, but is slightly smaller and darker. She builds her nest on the ground under a bush, among stones, or in a rocky cleft, and lines it with grass and a thick layer of down. She incubates the clutch of 6—12 greenish-white or pale grey eggs for 24 days. Then she watches over her young, guides them to food, and shelters them under her wings at night. This duck feeds on water plants, seeds, which are pecked up in the fields after nightfall, and on aquatic insects, crustaceans, molluscs and worms.

COTTON PYGMY GOOSE
Nettapus coromandelianus

The Cotton Pygmy Goose, which is only 33 cm long, is native to the whole of southern and south-eastern Asia, the Greater Sunda Islands, the Philippines and Australia. The female resembles the male, but is brown instead of green and the white parts of her plumage are spotted. This handsome goose can be seen in low-lying land, on rivers, lakes and deep-water swamps. For most of the year it lives in small groups of 5—15, though sometimes as many as 500

Ruddy Shelduck

Spotbill Duck

birds may assemble. In the breeding season, it forms pairs. Nesting takes place in July and August in northern India, between February and July in Sri Lanka, and in other months according to the area of distribution. The nest is built by the female in a hollow tree, usually 2—5 m above the water, but sometimes as high as 20 m above the water. The nesting cup is lined with grass and down. The clutch of 6—14 whitish eggs is incubated by the female for 26 days. As soon as the down on the newly hatched goslings has dried, they jump out of the nest, landing softly on the surface of the water. The Cotton Pygmy Goose feeds on seeds, grass, water vegetation and on rice grains in the paddy fields. Crustaceans, worms and water insects are occasionally eaten, mainly by the young.

Cotton Pygmy Goose

Lesser Whistling Duck

LESSER WHISTLING DUCK
Dendrocygna javanica

The Lesser Whistling Duck is widely distributed throughout southern Asia, from where it extends to southern China and across the western part of southeast Asia to the Greater Sunda Islands. It reaches a length of 42 cm and the sexes are similar in colour. This duck frequents large areas of water, lakes and swamps, often visiting rice paddies and coastal lagoons. It prefers localities where there is thick cover,

and tall trees in which to roost. It is gregarious, living in small groups of 10—20, though sometimes forming larger flocks. In the nesting season which varies with the area of distribution, pairs of whistling ducks gather to make small colonies. The nest is built in a large hole in a tree, and is lined with small twigs and grass. Nests abandoned by herons are sometimes used, and a nest may occasionally be built on the ground, among roots. The nesting cup is not normally lined with down. The female, possibly aided by the drake, incubates 7—12 ivory white eggs for about 22—24 days. This duck feeds on water plants, seeds and rice grains, and on small fishes, tadpoles, molluscs and worms.

BLACK KITE
Milvus migrans

The Black Kite is found throughout temperate and tropical Asia, Europe and Africa. This raptor is 65 cm long and has a wingspan of 120 cm. In the Himalayas it occurs at a height of 2 200 m. Kites of the more northerly regions are migratory, and those living in monsoon areas leave in the rainy season for drier localities. The Black Kite prefers to live near large rivers. Here it picks up dead fishes and catches other creatures in flight. It also feeds on the remains of pigs and other domestic animals. It is a common habit in the Asiatic tropics to throw domestic rubbish into the rivers, and scores of these graceful birds of prey often cruise above the water in the hope of finding food. The Black Kite is also abundant both inside and on the outskirts of towns. Here it feeds on refuse on derelict land. Flocks of kites are often seen in fishing villages, devouring fish entrails, which lie in heaps after the fishes have been gutted. They also capture small mammals, reptiles, frogs, molluscs and insects, and kill an occasional chicken.

In the breeding season the Black Kite lives in pairs. The nest is made of branches, roots, leaves and pieces of hide. The platform-like construction is situated in tall trees of the genera *Melia, Ficus, Dalbergia, Mangifera,* and often in the tops of palms such as *Cocos* or *Borassus.* The nest is normally 7—14 m above the ground, though sometimes nests are built as much as 30 m high. The female lays 2—4 eggs which vary in colour from grey and green to pink, and which have brownish-black, rust-coloured or red spots. She incubates the clutch for 28 days, the male only occasional-

Black Kite

ly relieving her. Both parents feed the chicks, which leave the nest when they are 42—45 days old.

BRAHMINY KITE
Haliastur indus

The Brahminy Kite is indigenous to southern Asia, southern China, Thailand, Vietnam and Annam. This raptor is about 48 cm long and frequents rivers, lakes and rice paddies. In the Himalayas it is found up to a height of 1 800 m. It often visits fishing settlements and harbours, where it scavenges dead floating fishes, or catches live ones. It also hunts frogs, crabs, lizards and snakes, small birds and insects. In villages it often seizes domestic poultry.
The nest is built of branches or palm leaves by both partners. It is situated 6—15 m above the ground in a large tree such as a fig tree, usually near water. The nesting cup is lined with plant wool and pieces of hide. The clutch of 2 greyish-white, russet-spotted eggs is incubated mainly by the female for 26—27 days. The young are fed by both parents.

GREY-HEADED FISHING EAGLE
Ichthyophaga ichthyaetus

The Grey-headed Fishing Eagle ranges from Pakistan across India, Burma and Malaysia to the Greater Sunda Islands, extending to Sulawesi and the Philippines. This eagle, which is about 75 cm long, mainly frequents rivers, or large mountain lakes in wooded areas, where it catches fishes in the water. After catching a large fish in shallow water, this bird wades with its heavy burden to the shore. It flies low above the water as it hunts. It also preys on small birds, especially fowl, and catches the occasional squirrel. This eagle lives singly or in permanent pairs. The nest, which is 1.5 m across and 1 m high, is made of branches and is situated in the tops of trees such as *Bombax*, *Terminalia* or *Albizzia*, near lakes or rivers. The nest is 10—30 m above the ground. Each pair builds 2—3 nests, 1—2 km from each other, and inhabits them in successive seasons. The nests are regularly repaired, and the birds choose one of them in which to sleep and rest throughout the year. The female lays 2, or occasionally as many as 4 white eggs, and sits on them alternately with her partner for 28—30 days. The young, which leave the nest after 10 weeks, are fed by both parents.

Grey-headed Fishing Eagle

Brahminy Kite

PHILIPPINE SERPENT EAGLE
Spilornis holospilus

The Philippine Serpent Eagle lives on river banks and lake shores of the Philippines. This raptor, which is about 70 cm long, is often seen perching on exposed horizontal branches above shallow water. It hunts small, slender fishes, frogs and crabs in the water and snakes and lizards in the trees. It occasionally catches small mammals and birds.

Adult eagles usually pair for life. Both partners work together to build the nest in a tall tree, about 15—25 m above the ground. The nest is made of branches and is lined with grass stalks. The female incubates 2 whitish eggs with black and brown spots for about 35 days. The young are fed by both parents.

SARUS CRANE
Grus antigone

The Sarus Crane is the largest of all the cranes. It reaches a length of 150 cm and a height of 190 cm, the female being slightly smaller. Its area of distribution covers northern and central India, Assam, Burma and Thailand, and it also occurs in the Philippines. It usually lives in pairs, and after the young have fledged the families stay together for some time. This crane prefers areas of extensive swampland and water, or seeks river banks and rice paddies. Cranes have conspicuous courtship behaviour. They make loud trumpeting sounds, accompanied by hopping, spreading of wings, and running around in graceful circles. The nest is built on a raised area in the middle of open marshland or swamps. Nests are also found in shallow water, the surrounding water being up to 1 m deep. A nest of this kind is often flooded during heavy rain. The female lays 2 spotted eggs and sits on them for 32 days, the male relieving her while she feeds. Both parents care for their young, helping them to find food. Young cranes are extremely active. As soon as they are dry, they begin to wander in the thick vegetation and they can swim skilfully. The Sarus Crane is omnivorous, feeding on green shoots, seeds and berries, and also catching insects, molluscs and small reptiles. It often raids cultivated fields for grain, fearless of men working nearby. In winter, it roams the countryside in small flocks, but each pair keeps its distance from the others.

Blue-breasted Banded Rail

BLUE-BREASTED BANDED RAIL
Rallus striatus

The Blue-breasted Banded Rail is distributed from India across south-eastern Asia to the Greater Sunda Islands, Sulawesi and the Philippines. It reaches a length of about 25 cm. It inhabits overgrown swamps, rice paddies and mangrove thickets. During the day, it sleeps, hidden in the vegetation. It forages at dusk and at night, walking or running through the marshland and hunting for insects, spiders and molluscs. It also eats green shoots and small seeds.

Each pair builds a carefully concealed nest made of dry and green plant material, among reed stalks or in tufts of bamboo. The female lays 4—8 spotted eggs which she incubates, helped by the male, for 20 days. As soon as they are dry, the newly hatched young scatter in the neighbourhood, being extremely adept at making their way through the thick tangle of plants. The adults bring them food in their beaks. This rail is a good swimmer, and when danger threatens is able to dive to safety.

WHITE-BREASTED WATERHEN
Amaurornis phoenicurus

The White-breasted Waterhen inhabits swamps, rice paddies and mangrove jungles throughout southern and south-eastern Asia, in the Greater Sunda Islands, Sulawesi and the Philippines. It can be found at a height of 1 300 m in the mountains, or even occasionally at 2 000 m. This waterhen reaches a length of 34 cm, and both sexes are identically coloured. At dusk, the birds make wailing sound. The more superstitious of the local people believe the sounds are the voices of the dead, and fear to walk in the swamps at night. The White-breasted Waterhen spends most of the day hidden in thick vegetation, but in the early morning and afternoon it comes out into the open to forage, carrying its short tail erect as it walks. It preys on insects, worms and spiders, and also feeds on seeds and green plant material.

Waterhens usually live in pairs which defend their own territories. The nest is built by both partners from leaves and stems, in dense vegetation in a swamp, or low above the water in the branches of a bush. The clutch of 4—7 creamy-bluish eggs with russet and greyish-purple spots is incubated by both birds for 20 days. The young leave the nest as soon as they have dried, and the adults bring them food in their bills for the first days after hatching.

WATERCOCK
Gallicrex cinerea

The Watercock ranges over the whole of southern and south-eastern Asia, the Greater Sunda Islands, Sulawesi, and the Philippines. This bird, which is about 42 cm long, lives in swamps containing large expanses of water, and in rice paddies. Outside the nesting season it flies to coastal areas, and dwells in lagoons and over freshwater lakes near the sea. It is a shy bird, and remains always in the shelter of undergrowth or tall clumps of vegetation.

Pairs of watercocks arrive at their nesting grounds in the monsoon season, and take up occupation of territories which they fiercely defend against intruders. They build nests among the thick tangle of vegetation, using large quantities of leaves and grass. The female lays 2 eggs, cream or reddish-white in colour, with large, russet, red or violet spots. Both partners incubate the clutch for 22 days and then feed the young. The Watercock feeds on seeds, the leaves of aquatic and swamp plants, insects, spiders, worms and molluscs.

White-breasted Waterhen

Watercock

PURPLE SWAMPHEN
Porphyrio porphyrio

The Purple Swamphen has an extensive area of distribution, ranging from southern and south-eastern Asia and the Greater Sunda Islands to Australia, Africa and southern Europe. It frequents large swamps containing pools of water or quiet backwaters with overgrown banks. It forages in the early morning, and during the day it sometimes walks on the huge, floating leaves of water plants, using its extremely long toes. It is adept at making its way through the dense cover of swamp plants. Swamphens feed on seeds, green leaves, insects and other invertebrates, and have been seen to eat the eggs of other birds.

A large nest made of plant material, and often situated among flowering lotuses, is built by both partners. They then sit on the clutch of 4—8 greyish-yellow eggs with brown and violet spots for 22 days. The young are black all over, except for a red spot on the chest and red feet. They stay in the nest for about 3 days, after which they can be seen swimming or running nimbly on the leaves of water plants. At first they peck food from the bills of their parents, but later they forage for themselves, while the adult birds guide and protect them.

MASKED FINFOOT
Heliopais personata

The Masked Finfoot is native to India, south-eastern Asia and Sumatra. It attains a length of 55 cm. The female resembles the male in colour, but the centre of her throat and the front of her neck are white instead of black. This finfoot frequents overgrown lakes, pools and mangrove jungles, each pair defending its own territory. Finfoots often build a floating nest made of pieces of aquatic and swamp plants. Both partners incubate the clutch of 4—6 spotted eggs for 22 days and then tend their offspring. The diet of the Masked Finfoot consists of small invertebrates, green plants and seeds.

YELLOW-WATTLED LAPWING
Vanellus malabaricus

The Yellow-wattled Lapwing, which is about 30 cm long, is native to Sri Lanka, where it inhabits moist

Purple Swamphen

meadows and fields, as well as stone and grass-covered sites. Outside the nesting season it wanders in small flocks, which in May and June break up into pairs. Courtship takes place on the ground, where the partners run around each other, and in the air, where they make acrobatic nuptial flights. Lapwings line a shallow depression among stones or in tufts of grass with stalks and leaves. There the female lays 4 eggs, greyish-yellow with grey and rust-coloured spots, which blend so perfectly with their background that the nest can scarcely be seen. The clutch is incubated by both partners for about 26 days. The day after hatching, the young scatter in the neighbourhood, and they begin to fly when they are 5 weeks old. This lapwing feeds on insects, insect larvae, spiders, worms, molluscs, seeds and green plants.

RED-WATTLED LAPWING
Vanellus indicus

The Red-wattled Lapwing is found from the Caspian Sea to south-western China, India, Sri Lanka, Malaysia and Sumatra. It reaches a length of 34 cm and both sexes are identically coloured. Its habitats include swamps, river banks and rice paddies, although it also occurs in thin, humid woodland and in meadows. This lapwing usually lives in pairs, and forms family groups after the fledging of the young. It preys on insects, spiders, worms and molluscs in the grass, often flying some distance to forage in drier localities. Its flight is slow but skilful.

The nest is built on the ground in a depression sheltered by large stones or by a bush. The birds line it sparsely with grass stalks and leaves. Both partners incubate the clutch of 4 greyish eggs with black spots for 25 days. Young lapwings are small but their heads are large in comparison with the rest of the body, and they have long legs. They are spotted black and white.

SOCIABLE PLOVER
Chettusia gregaria

The Sociable Plover, which is about 33 cm long, is native to an area stretching from eastern Europe to central Asia. In autumn flocks of these plovers combine to form large groups, often in company with other related birds, to migrate to their winter grounds in India and Sri Lanka. They return to their homeland in

Sociable Plover

5 weeks. The diet of these birds consists of insects and insect larvae, worms and molluscs.

PAINTED SNIPE
Rostratula benghalensis

The Painted Snipe is a beautiful bird which inhabits Sri Lanka, India, southern China, Thailand, the Greater Sunda Islands and the Philippines. The female attains a length of 28 cm, the male measuring only about 26 cm. She is also more colourful than her mate. This may be because the male is the nest builder, while the female defends the territory. The clutch of 4 pinkish, brown-spotted eggs is incubated solely by the male for about 3 weeks, and the young are protected by him. Some ornithologists maintain that each female has several mates, but this has not been confirmed. The nest is made of stalks and leaves, and it is situated on the ground in a tuft of grass, usually in a swamp. Nesting takes place between November and May. At this time the female makes loud noises after nightfall, while the male is virtually silent. When the brood is fledged, families merge to form small flocks, those from the more northerly regions migrating to Malaysia. This snipe flies only slowly. During the day, the birds hide in thick vegetation, becoming active between sunset and dawn, when they forage in swampland and shallow water for aquatic insects, insect larvae, worms, small molluscs, and seeds, especially rice grains.

March, and soon after arrival pairs occupy their nesting territories. The nest is built on the ground, in a depression lined with grass stalks, and the clutch of 4 spotted eggs is incubated by both partners for 25 days. When their down has dried, usually the day after hatching, the young leave the nest to wander in the neighbourhood. They hide in tufts of grass or among stones, and their coloration helps them to blend with their background and so escape their enemies. The young plovers are ready to fly after

PINTAIL SNIPE
Gallinago stenura

The Pintail Snipe, which is about 25 cm long, is indigenous to an area from eastern Asia north to the Himalayas. It leaves its homeland in September to migrate to India, Sri Lanka, the Greater Sunda Islands, Sulawesi and the Philippines. It returns to its nesting grounds in the swamps, rice paddies and grassy localities in March. The nest is built in a tuft of grass or among other plants, and is lined with dry leaves and stalks. The female lays 4 eggs with olive and brown spots, and incubates them for 19—21 days. The newly hatched young soon scatter in the surrounding tall vegetation. They are cared for by both parents, who bring them food for the first few days after hatching. The young snipes begin to fly after 3 weeks, and in

Painted Snipe

autumn the families merge to form flocks. The Pintail Snipe feeds on insects, spiders, worms, and molluscs which the birds peck out with their long beaks from the soft, muddy soil. They forage mainly after nightfall, only occasionally hunting in the early morning or late afternoon.

Pintail Snipe

PHEASANT-TAILED JAÇANA
Hydrophasianus chirurgus

The Pheasant-tailed Jaçana is distributed throughout south and south-east Asia, the Greater Sunda Islands and the Philippines. It reaches a length of 30 cm, but in the breeding season, the males grow elongated tail feathers up to 25 cm long. The Pheasant-tailed Jaçana is a denizen of swamps, marshes and lotus lakes. In many areas it lives in flocks, which walk on the large floating leaves. Jaçanas are good swimmers, although they are rarely seen in open areas of water. They usually prefer to stay among dense water and swamp vegetation. Jaçanas are slow but efficient fliers, and in flight their white wings contrast with their black underparts. The Pheasant-tailed Jaçana feeds on water plants, seeds and small invertebrates such as insects, worms and molluscs.

During the courtship season the male attracts the female by a loud call, at the same time raising his elongated tail feathers. Both partners bring pieces of plant material to make a heap which forms the foundation of their nest. They build the nest among lotus leaves, and the female occasionally lays an egg directly on a large leaf. The clutch averages 3—4 pale bronze-coloured eggs with dark dots and spots. The male incubates them for 24 days and watches over the young, while the female joins another male. In one season, each female may produce as many as 10 clutches. Young jaçanas are able to fly after 6 weeks, by which time the male is often sitting on another clutch.

BRONZE-WINGED JAÇANA
Metopidius indicus

The Bronze-winged Jaçana is distributed from India to south-eastern Asia, Sumatra and Java. It reaches a length of 30 cm and has extremely long toes, adapted for running over the broad floating leaves of aquatic vegetation. Both sexes are the same colour. This jaçana mainly inhabits swamps containing extensive areas of water, or lakes overgrown with lotus plants. It often moves over the surface in small flocks, and is

Pheasant-tailed Jaçana

Bronze-winged Jaçana

White-winged Black Tern

ter plants. The female usually lays 3 yellow eggs with brown and black lines and spots. The male incubates the clutch for 23 days and later rears the young.

a good swimmer and diver. It flies only clumsily and seldom flies for any distance. Jaçanas feed on small invertebrates, mainly aquatic species, and on seeds and green plants.

Each female has several mates, each of which builds a flat, floating nest of stalks and leaves among the wa-

WHITE-WINGED BLACK TERN
Chlidonias leucoptera

The White-winged Black Tern, which reaches a length of 25 cm, inhabits lakes, swamps, river deltas and rice paddies of eastern Asia and across a large part of temperate Asia. It overwinters in the tropical areas of Sri Lanka, Burma, Malaysia, the Greater Sunda Islands, and even Australia, returning to its native land in late April or early May. This tern usually lives in pairs which join to colonies. The nest is built on bent reeds in shallow water or on floating islands of vegetation. It is constructed by both partners from pieces of aquatic plant material. The female usually lays 3 spotted eggs, variable in colour, and sits on them alternately with the male for about 16 days. Both parents feed the young on insects and small fishes, which they catch on the wing.

Malaysian Fish Owl

MALAYSIAN FISH OWL
Ketupa ketupu

The Malaysian Fish Owl is indigenous to south-eastern Asia and the Greater Sunda Islands. It stands up to 50 cm tall. It is a resident of woods, near rivers, lakes and rice paddies. This owl lives in pairs, but each partner has a regular site in which to rest and sleep during the day, usually in the dense branches of a tree, or in a large hole in a tree. As soon as darkness falls, the owl leaves its shelter to hunt. It perches on a dead branch above the water, or finds a vantage point on a high bank, or on a large stone in the shallows. It preys on fishes and crabs, and on land it catches lizards, snakes, small rodents and large insects.

The nest is built by both partners in a hole in a large tree, or in a small cave, usually above the water. It is not lined except for a few scraps of skin from the prey. The female lays 2—3 white eggs and sits on them for 5 weeks, while the male brings food which he passes to her near the nest. The male feeds the whole family after the young have hatched, while the female stays on guard and takes charge of food brought by the male. In due course she joins him in hunting. Young fish owls begin to fly when they are 10 weeks old.

BROWN FISH OWL
Ketupa zeylonensis

The Brown Fish Owl is about 55 cm long. It is widespread from the Middle East to southern China and south-east Asia, and it is also common in India and Sri Lanka. This fish owl is a resident of low-lying land, although it can be seen at a height of 2 000 m. It lives beside forest streams and lakes, and visits rice paddies, where it hunts for fishes and crustaceans, and catches reptiles on the banks. In the nesting season it makes a variety of deep, often harsh noises. The nest is built in the hollow of a large tree, such as a fig or mango, or in a rock cleft. The same nesting site is usually occupied for many successive seasons. The female lays 1—2 eggs and incubates them for 35 days, the male bringing food for her to a place close to the nest. The breeding season varies with the area of distribution, taking place in January and February in northern India, and in December in the south.

THREE-TOED FOREST KINGFISHER
Ceyx erithacus

The Three-toed Forest Kingfisher has an area of distribution covering south and south-east Asia, Sumatra and Borneo. This species is 14 cm long. It frequents forest streams and lakes, mainly in low-lying land, and only occasionally dwelling in hills up to a height of 600 m. It usually rests on dead branches above water, from which it can easily take off and swoop after its prey. It also flies noisily along streams, in its search for tiny fishes, freshwater crabs, and the occasional small frog.

The nest is built in a bank above the water, the birds

Brown Fish Owl

excavating a burrow up to 60 cm deep and 5 cm in diameter. The female lays 2—3 pure white eggs in the nesting chamber, which is 15 cm across. She incubates the clutch for 18 days, while the male feeds her, and he occasionally also sits on the eggs. The young are fed by both parents and leave the nest after 3 weeks.

BLUE-EARED KINGFISHER
Alcedo meninting

The Blue-eared Kingfisher is widespread throughout south and south-east Asia, the Greater Sunda Islands and the Philippines. It is a small species, reaching

Three-toed Forest Kingfisher

Blue-eared Kingfisher

Stork-billed Kingfisher

a length of about 15 cm. It inhabits forest streams and lakes, but it can also be encountered in rice paddies near woodland. Outside the nesting season it wanders, usually singly, in the neighbourhood of its home, and it forms pairs to breed. Kingfishers dig corridors about 80 cm long in clay or sandy banks, terminated by a nesting chamber. They use their feet to throw out the soil. The female sits on the clutch of 4—7 white eggs for 20 days, occasionally being relieved by the male, who brings food to her. The adults feed their young for 25 days in the nest and for a further 2 weeks after fledging. When hunting, kingfishers often dive.

STORK-BILLED KINGFISHER
Pelargopsis capensis

The Stork-billed Kingfisher, which is up to 40 cm long, has a characteristic, massive, red bill. This bird is native to south and south-east Asia, the Nicobar Islands, the Sunda Islands, Sulawesi and the Philippines. Here it lives along forest streams and lakes, in rice paddies, and in the coastal mangrove swamps. Although it prefers lowlands, it also occurs in mountainous areas up to a height of 1 200 m. Its favourite vantage point is a strong, shady branch above the water, from where the kingfisher can keep watch for its prey. It mainly hunts fishes, but it also catches frogs, crabs and other small creatures. The prey is usually swallowed whole, head first.

The nesting hole is generally excavated in the bank of a river or lake, though a pair of kingfishers has been observed to carve out a hollow in the trunk of a large, dead tree in which the wood had rotted. The female lays 2—5 white eggs which she incubates for 24 days, the male occasionally helping her. The young are fed for about 25 days in the nest by both parents, and for a further 10 days after fledging.

WHITE-BREASTED KINGFISHER
Halcyon smyrnensis

The White-breasted Kingfisher is very common in an area extending from the Middle East to southern China, south-eastern Asia and the Philippines. It reaches a length of about 28 cm. This kingfisher prefers open country near unpolluted streams and lakes. In the mountains, it occurs up to a height of 1 600 m, and it can often be seen in city parks and gardens with small lakes or reservoirs. The White-breasted Kingfisher hunts grasshoppers, frogs and small lizards on land, but is adept at catching fishes as well. During most of the year it is solitary, each individual occupying its own hunting range.

Nesting takes place between December and June, but kingfishers usually form pairs in March or April. The nest is excavated in a bank above or near to water. The tunnel may be as long as 2 m, but it is usually about 1 m long. Kingfishers sometimes nest in a hollow tree, and a pair has been seen to try to peck out a hole in a wall, though without success. The female lays 3—5 white eggs, which darken during incubation. The young hatch out after 25 days, incubation being carried out mainly by the female. The nestlings are fed on insects by both parents.

BLACK-CAPPED KINGFISHER
Halcyon pileata

The Black-capped Kingfisher inhabits the whole of south-eastern Asia, extending to Korea, the Philippines, Sulawesi, India and Sri Lanka. It is about 30 cm long. Its habitats include waters along the edges of forests, coastal mangrove jungles, lagoons, and sometimes rice paddies. It often perches on bamboo stalks leaning over the water. This kingfisher preys mainly on freshwater crabs, easily cracking the carapaces with its strong beak. Outside the nesting season the Black-capped Kingfisher roams the countryside, often flying hundreds of kilometres from its home. Its breeding habits are similar to those of the White-breasted Kingfisher.

CHINESE ALLIGATOR
Alligator sinensis

The Chinese Alligator inhabits the lower reaches of the Yangtze River in eastern China. It attains a length of 175 cm, and its feet are not webbed. The existence of the Chinese Alligator was only revealed to the rest of the world in 1879, although it had been known in China for a long time. It is mentioned in books dating from several centuries before, and in illustrations it was considered to be related to an imaginary dragon. The Chinese Alligator is not very common in its native land, though it is not rare either. However, catch-

White-breasted Kingfisher

ing these reptiles in such an extensive territory is difficult, particularly as they are very cautious, so their price to zoos and wildlife parks is high. Hunters set out to catch these alligators in autumn, when they crawl ashore to dig out holes in which to hibernate. The hunters use dogs to track down the alligators in their shallow burrows, and some alligators are

Black-capped Kingfisher

trapped in fishing nets. The diet of this species consists of fishes, molluscs and crustaceans, while young alligators also feed on insects and spiders. The breeding habits of the Chinese Alligator remain largely unknown.

MUGGER or MARSH CROCODILE
Crocodylus palustris

The Mugger or Marsh Crocodile is found in Sri Lanka, India, Pakistan and Indonesia. It is a freshwater species, occurring in lakes and quiet backwaters. Although it may exceed 4 m in length, such large specimens are rare, and the average length is 3 m. In many regions, particularly in India, the Marsh Crocodile is a sacred animal. It may not be disturbed or harmed, and there are even crocodile temples complete with ponds, where the reptiles are kept and fed on goat meat by specially trained priests. The Marsh Crocodile is not usually hostile to man, but there are a few recorded cases of large specimens killing men.

The Marsh Crocodile feeds predominantly on fishes, but it also catches birds and small mammals as they drink at the river's edge. The female builds a large heap of leaves, mud, sticks and other plant materials on the bank close to the water and here she lays several scores of hard-shelled eggs. Then she remains on guard to chase away predators, such as monitors. After 10 weeks or so, when she hears the peeping sounds made by the young before they hatch, she pushes the heap away to help the young crocodiles to crawl out.

FALSE GAVIAL
Tomistoma schlegeli

The False Gavial inhabits Malaysia and Borneo. It can reach a length in excess of 5 m, but very large specimens are rare. This crocodile lives in rivers with banks overgrown with vegetation. It prefers quiet coves, where it often lies motionless near the surface of the water, and it climbs on small floating islands. It

Mugger or Marsh Crocodile

False Gavial

feeds mainly on fishes, but it also catches amphibians and birds, together with mammals which venture into the shallows to drink. The female lays some scores of eggs in holes dug out in the bank near the water, and watches over the clutch for 9—10 weeks until the young hatch out. They are immediately able to fend for themselves, but they remain in shallow water close to the bank, hiding among the water plants. The young crocodiles feed on insects, spiders and tiny fishes.

INDIAN GAVIAL
Gavialis gangeticus

The Indian Gavial is native to India, where it lives in the rivers Sind, Ganges, Mahanadi and Brahmaputra. It is also said to have been seen in Burma. With a length of 7 m, it is one of the largest species of crocodile, but despite its size, it is not aggressive, and is not known to have ever attacked man. The narrow and elongated jaws of the Indian Gavial are adapted for fishing. It is a sacred animal almost everywhere in the area of its distribution, and as such may not be killed or disturbed. Furthermore the catching of young gavials for transfer to zoos has always to be authorized. Evolutionarily, the Indian Gavial is an ancient animal. More than 60 million years ago it occurred in India, Africa, Europe and North America,

but the area of distribution has since diminished. It lives mainly in rivers, though it may sometimes be encountered in lakes or river deltas. Its jaws are rimmed with at least a hundred sharp, pointed teeth, so any fishes trapped in the gavial's mouth cannot escape. Like other crocodiles, this species also catches fishes by rapidly jerking its head sideways, sometimes seizing several fishes at a time. It can handle prey admirably with its jaws without letting it go. The Indian Gavial was for a long time believed to eat nothing but fishes, but the stomach contents were investigated and they revealed the remains of dogs, birds, goats and jewellery. The presence of the jewellery demonstrates the habit shared with other species of crocodile, of swallowing stones and various other small objects. The female lays about 40 eggs in a hole in a sandbank, larger and older females producing bigger clutches. The eggs measure 9×7 cm and have hard shells. The young hatch out after 10 weeks.

INDIAN SOFT-SHELLED TURTLE
Trionyx gangeticus

The Indian Soft-shelled Turtle is a denizen of the rivers of India, occurring predominantly in the

Indian Gavial

Indian Soft-shelled Turtle

Ganges basin. It reaches a length of over 40 cm and has characteristic sharp jaws covered by fleshy lips. It spends most of its time in the water, occasionally crawling on to land, though never venturing far from water. It sometimes buries itself in the sand to conceal itself from its enemies. The female produces 2—3 clutches a year, each consisting of a score or so of eggs. She lays them in holes in the ground and carefully covers them. Young soft-shells hatch out after 45—60 days, and immediately scatter among the vegetation in the shallows. The young feed on insects, molluscs and crustaceans, while the adults hunt for fishes. They lie in wait for hours until the prey approaches their jaws, and then they pounce on the fish with great speed. This turtle can remain submerged for 15 minutes without breathing, but it usually lies in the shallows with its elongated snout above the water so that it can breathe comfortably.

THREE-LINED BOX TURTLE
Cuora trifasciata

The Three-lined Box Turtle is found in Vietnam and southern China, and is easily recognized by the three black longitudinal stripes on the 25 cm-long carapace.

This turtle lives in pools, quiet stretches of rivers and flooded rice paddies. It is abundant in some localities, but it is a cautious animal, and is therefore rarely seen in the wild. It frequents the banks of rivers where it basks in the sun, but the slightest vibration in the ground, such as would be caused by a man walking, is enough to warn the turtle, which immediately slides into the water. It is sometimes caught in the nets of fishermen. The diet of this turtle consists of small aquatic invertebrates, small fishes and pieces of plant material. The female lays her eggs on land, in holes which she digs out in the sand.

ROOFED TERRAPIN
Kachuga tectum

The Roofed Terrapin is one of the most beautifully coloured turtles. It lives in northern India and western Pakistan. It reaches a length of 25 cm, and has a characteristic, tall, roof-shaped carapace. The underparts are orange in colour, or red with black spots. This turtle only frequents unpolluted waters, where it feeds predominantly on green water plants, though occasionally catching insects and their larvae living in the water, or molluscs and tiny crustaceans. The young are more carnivorous. These turtles often bask on the river banks, or climb on to floating pieces of wood. Here they are safer from their enemies and they can easily slip into the water. The female lays just a few eggs in holes dug in a sandy bank. The young hatch out after 9 weeks and are heavily preyed upon by raptors, herons and storks.

BIG-HEADED TURTLE
Platysternon megacephalum

The Big-headed Turtle is distributed in south-east Asia and southern China, where it lives in small, fast-

Three-lined Box Turtle

running mountain streams. It has a carapace up to 40 cm long, and a huge head. The carapace is flattened to reduce resistance to the current, while its powerful claws grip the stony bottom for support. The Big-headed Turtle feeds mainly on freshwater crustaceans and gastropods, crushing the shells and carapaces with its strong jaws. When hunted, this turtle has to be caught by the tail, for it can inflict serious injuries. This species is very common in some places, but it is not easily caught in the mountain streams. The natives use nets stretched across the water and weighted at the bottom. They enter the water a little further upstream, forcing all the water creatures towards the nest, and trapping the turtles. Big-headed turtles often climb on to stony banks to bask in the sun, but they always stay close to the water. The female lays 2 eggs in a depression dug out on the land.

EAST INDIAN WATER LIZARD
Hydrosaurus amboinensis

The East Indian Water Lizard is confined to the Molucca Islands. It can exceed 1 m in length, and has a thick tail with a tall leathery crest on the upper part, strengthened by long bony processes from the tail vertebrae. This water lizard lives among trees and bushes beside rivers and lakes, where it feeds on shoots, berries and fruit. It also eats water plants, insects, spiders and centipedes. When it is in danger, it dives into the water, for it is an excellent swimmer. In this way it often escapes predators such as large raptors or tree carnivores. The female buries her soft-shelled eggs in the sand on the river banks. The natives collect the eggs and eat them, and they also catch the lizards, for they are regarded as a delicacy.

WATER MONITOR
Varanus salvator

The Water Monitor is found in south-eastern Asia, ranging in the north as far as southern China and also occurring in Sri Lanka, Indonesia, and occasionally in the Molucca Islands and the Philippines. It can grow to a length in excess of 3 m, the largest one to be caught having measured 3.21 m. The Water Monitor lives on the banks of rivers or small lakes, for water is its element. It is an excellent swimmer and diver, and hunts in the water for fishes and other aquatic animals, being able to remain underwater for more than half an hour at a time. It is equally nimble on land, moving speedily in the scrubland, where it is difficult to catch. These monitors live among bushes and low trees whose branches overhang the river, so that when danger threatens, they can jump into the water. They also use the branches as a vantage point when preparing to pounce on their prey. Under cover of darkness, these monitors take eggs, chickens, hens, ducks and

Big-headed Turtle

East Indian Water Lizard

geese from human settlements, so the local inhabitants pursue them ruthlessly. When a monitor is unable to escape, it defends itself vigorously, using its tail to hit its assailant, and also biting and scratching. In some regions, monitors are caught by tribal magicians. They tie them up near a bonfire and provoke them to blow on the flames, while they brew their poisons. In this way, they aim to make the brew more lethal, for many believe that the Water Monitor is a dangerous animal. Various 'medicines' are also made from its fat. This species once provided — and in some places still does — an important component in the poison-making ritual. The Singhalese produce a most fearsome mixture of poisons by catching a great number of venomous snakes, making incisions in their heads and suspending them above a vessel into which the poison is believed to drip. They mix the snake blood with arsenic and other strong poisons, and brew the liquid in a human skull. Then they tie several water monitors around the bonfire and torture them until the animals begin to hiss towards the flames. Of course, the efficiency of this mixture depends upon the arsenic. In Vietnam, these monitors are caught for their attractive and durable skin, which is used to make handbags, shoes and ornaments. Hunters dig out the monitors from burrows up to 2 m deep in the river banks.

The Water Monitor feeds on small vertebrates including other kinds of lizards, and it also catches freshwater crabs. It is itself preyed upon mainly by large

Water Monitor

Chinese Water Dragon

pythons. In summer the female lays 5—25 eggs in holes dug out in the sand or in soft soil. The young monitors feed on insects and other invertebrates at first, and later they catch small rodents and birds.

CHINESE WATER DRAGON
Physignathus concincinus

The Chinese Water Dragon is a resident of overgrown banks of rivers and lakes in southern areas of southeast Asia. It can grow to more than 1 m in length. This water dragon stays always on branches over the water, so that it can jump when danger threatens. Here its flattened tail enables it to swim skilfully. It picks juicy berries and fruit in the trees and bushes, and it also collects fallen fruits. It also feeds on insects, frogs and small lizards. The female lays 15—20 eggs in a depression which she digs in the ground, and then she leaves the clutch. The newly hatched young catch insects and other tiny invertebrates, and nibble soft fruit.

RETICULATED PYTHON
Python reticulatus

The Reticulated Python is one of the longest of the snakes, and is found in southern China, Vietnam, Burma, Malaysia and the Philippines. It inhabits regions where there is a wealth of water, and it is always found in the vicinity of rivers, lakes and rice paddies. It hunts after sunset, catching rats, other small mammals and birds. The larger specimens, which are over 7 m long, can devour mammals weighing 30 kg or more. Indeed in Malaysia, a python measuring 8 m seized and swallowed a pig which weighed 72 kg. This is the largest prey known to be taken by a snake. In theory a huge python could devour a man, or certainly a child, but pythons, like other snakes, usually avoid man. There is an account of a reticulated python suffocating and swallowing a fourteen-year old boy on the island of Salebabu in Malaysia, but such cases are very rare.

Malaysians have a special method of hunting large reticulated pythons. They make a strong cage of bamboo sticks and use a small live pig as bait. During the night a foraging python crawls into the cage, suffocates the piglet in the coils of its body and swallows it whole. The prey is so big that the snake's stomach becomes distended, and the python can no longer get out of the cage. In the morning, the hunters are able to take the snake alive or kill it. The skin is used to make various products such as purses and shoes.

In the past some people have maintained that a python can live to be five hundred years old, but this is not so. The life span of the Reticulated Python averages 30 years.

The female lays clutches of 20—100 eggs protected by leathery shells. The eggs are 8 cm long and 5 cm wide. She envelopes the clutch in the coils of her body, guarding and warming the eggs. At an average temperature of 28° C the young hatch after about 80 days. They are 30—50 cm long at birth. While she watches over her clutch, the female usually takes no food, but occasionally drinks at a water-hole.

Reticulated Python

BLOOD PYTHON
Python curtus

The Blood Python is native to Malaysia, Sumatra and
Borneo, and it is sometimes found in southern Vietnam. It is a small species reaching a length of only
1.5 m, and it has a stout, clumsy body. The Blood
Python makes its home in swamps, where it catches
rats and other small rodents. When cornered it
charges violently at its assailant, which is remarkable
having regard to the thickness of its body. Its attack is
almost like a leap. This snake is easy to catch and its
beautiful, high-quality skin is used in the manufacture
of various products, although the rainbow colours
fade in processed skins. The female lays her eggs in
a heap and surrounds the clutch with her body sometimes for many weeks, in the same way as do other
pythons.

Blood Python

TENTACLED WATER SNAKE
Erpeton tentaculatum

The Tentacled Water Snake inhabits Thailand and
northern Malaysia. It grows to a length of 1 m, and
has two scaly outgrowths resembling antennae on its
snout. It has large venom teeth in the back of the upper jaw, which it uses to kill prey, consisting mainly of
fishes. This snake spends all of its life in water. The
worm-like movements of the appendages on the
snake's snout lure fishes towards its jaws. So far little
is known about the reproduction of this species.

STICKY CAECILIAN
Ichthyophis glutinosus

The Sticky Caecilian is a resident of moist localities
beside the rivers of Sri Lanka and southern India, extending to Malaysia and the Greater Sunda Islands.
This amphibian measures up to 40 cm in length, and
has no limbs. The female digs underground burrows
near to the water, where she lays about 12 eggs joined
together by a gelatinous string. She then coils round
the clutch to protect her eggs. When the larvae hatch,
they immediately crawl into the nearest water. They
lack external gills, and frequently go to the surface to
breathe. While living in the water, the larvae grow to
a length of about 17 cm. They have well-developed
eyes and flattened tails. After metamorphosis they return to land, and live in moist soil, where they feed on
worms and other tiny invertebrates.

INDIAN BULLFROG
Rana tigrina

The Indian Bullfrog is native to southern and south-eastern Asia. It is about 15 cm long, and is one of the larger species of frog. It abounds in rice paddies and frequents both rivers and lakes. It usually catches insects, spiders, worms, tiny fishes and tadpoles, while a large specimen can seize and swallow the young of lizards and snakes. The female lays over 5 000 eggs. Tadpoles hatch out after a week and undergo metamorphosis within 2 months. Young frogs, in particular, fall prey in great numbers to wading birds, raptors, owls, kingfishers and small carnivores. These large frogs are hunted locally, for their flesh.

SUMATRA BARB
Puntius tetrazona

The Sumatra Barb is widely distributed in several subspecies in Sumatra, Borneo, Malaysia, Thailand, southern Burma, India and Sri Lanka, the subspecies differing in coloration and striping. This fish reaches a length of 6—7 cm, the male being more slender and having a red-edged mouth. This barb lives in shoals, in running water in both lowlands and hills. At spawning time the female lays a very large number of eggs. The fry hatch out 24 hours after spawning and for about 3 days they live on their yolk sacs. During this time, they remain suspended on plants or stones, before beginning to swim and feed on tiny worms and crustaceans.

IRIDESCENT BARB
Puntius oligolepis

The Iridescent Barb is only about 5 cm long, and is native to Sumatran waters. The bright red male is more slender than the ochre yellow female. These barbs live in small shoals close to the muddy bottom, where they dig into the mud to find small worms. They also feed on the larvae of various insects, particularly gnats, and on crustaceans. The fry hatch 24 hours after spawning has taken place, and hang for about a week on water plants, until they exhaust their yolk sacs. After this they begin to swim. At the age of 6 months, young barbs have assumed their adult coloration.

RED BARB
Puntius conchonius

The Red Barb is found in northern India, Bengal and Assam. It attains a length of 14 cm. The female is predominantly yellowish-olive in colour. This fish lives in shallow waters which are densely overgrown with aquatic vegetation, and it spawns among the weeds. The development of the fry is similar to that of the previous species.

Sticky Caecilian

Tentacled Water Snake

Indian Bullfrog

tangle of water plants. Young goldfishes are dark immediately after hatching, and only assume their reddish or golden coloration after several months. Some specimens remain dark in colour, while others are spotted. The fry are rapacious feeders, eagerly pursuing insect larvae, worms and tiny crustaceans, though they also eat plant material.

GOLDFISH
Carassius auratus

The Goldfish is one of the best-known aquarium fishes, and was the first species of this kind to be bred in captivity. In Japan and China it was domesticated a thousand years ago. The Chinese have bred many colourful forms, such as pure white, yellow, yellow spotted, as well as forms with extremely elongated fins, and varieties with bizarre head shapes or huge eyes. The Goldfish was introduced in Europe, being first imported to Portugal in the 17th century.
The Goldfish inhabits still, overgrown water and slow-flowing rivers. In the spawning period, the female lays several thousand eggs among the dense

RED RASBORA
Rasbora heteromorpha

The Red Rasbora, which is about 4 cm long, is a denizen of flooded areas and streams in the lowlands of Thailand, the Malay Peninsula and Sumatra. The male is more lightly built than the female, and his dark spot reaches to the ventral fin. This fish lives in shoals but forms pairs before spawning. The female turns upside down to lay her eggs, attaching them to the underside of the leaves of water plants, where the male fertilizes them. The fry hatch after 24—30 hours and begin to swim 3—5 days later. The diet of the Red Rasbora consists of larvae, worms and tiny crustaceans.

SPOTTED RASBORA
Rasbora maculata

The Spotted Rasbora is one of the smallest species of fish, reaching a length of only 2—2.5 cm. It is native to the Malay Peninsula and Sumatra, where it is found in small shoals among the thick tangle of water vegetation. The female deposits her eggs on aquatic plants. The fry hatch out after about 26 hours and hang motionless on the leaves for about 4 days while

Sumatra Barb

Iridescent Barb

Red Barb

Goldfish

living on their yolk sacs. Then they scatter in the surrounding area. This species feeds mainly on plankton.

THREE-LINE RASBORA
Rasbora trilineata

The Three-line Rasbora inhabits the Malay Peninsula, Sumatra and Borneo. It is up to 15 cm long. This species seeks still and slow-moving waters with thick vegetation, and lives in large shoals. At spawning time the female lays her eggs on the leaves of water plants. The fry hatch after one day and remain attached to the plants for 3—4 days, until they exhaust their yolk sacs. The diet of this fish is composed chiefly of worms and insect larvae.

GIANT DANIO
Danio malabaricus

The Giant Danio is a beautifully coloured fish, about 12—15 cm long, found in the waters of south-western India and Sri Lanka. Despite its restricted area of dis-

Red Rasbora

Spotted Rasbora

Three-line Rasbora

225

tribution, it is extremely common in its native habitat. Danios move in large shoals in places densely overgrown with aquatic vegetation. The female spawns in the tangle of fine weeds. The fry hatch out 20—36 hours after spawning and live for 2—3 days on their yolk sacs, before scattering and hunting for their own prey. The Giant Danio feeds on insect larvae, worms and crustaceans, and on the leaves of water plants.

ZEBRA DANIO
Brachydanio rerio

The Zebra Danio is confined to still and sluggish waters of eastern India. It measures about 5 cm, and lives in small shoals among aquatic vegetation, where spawning also takes place. The fry hatch after about 30 hours, and 48—72 hours later the young begin to swim. The Zebra Danio feeds mainly on small insects, worms, crustaceans, and on the spawn of other species of fish.

PEARL DANIO
Brachydanio albolineatus

The Pearl Danio reaches a length of 6 cm. It has a characteristic mouth, directed almost upwards, which enables the fish to collect food at the surface. This attractive and undemanding species is native to the waters of south-eastern Asia and Sumatra. The Pearl Danio lives in shoals among water plants and close to the surface, where spawning also takes place. The fry hatch after 3 days, and for 2 days they remain

hanging on the water plants. The fishes stay near the surface and feed on tiny insects and plankton. Golden and blue forms have been bred in captivity.

SPOTTED DANIO
Brachydanio nigrofasciatus

The Spotted Danio is found only in Burma. This species, which is about 4 cm long, frequents rivers, where it lives along the banks, and still pools. The female lays 60—100 eggs among dense weeds, and the fry hatch out 30 hours after spawning. The diet of this danio is similar to that of the previous species.

RED-TAILED BLACK 'SHARK'
Labeo bicolor

The Red-tailed Black 'Shark' is a remarkably coloured, popular aquarium fish. It has a rich, velvety, black coloration with a bluish sheen, but its caudal fin is a shining orange-red colour. This species, which is 12—15 cm long, is native to Thailand. It lives in streams, near the banks, among the roots of water plants. It is extremely pugnacious and defends its territory against all other fishes. It feeds mainly on green plants, but insect larvae, worms and molluscs are also

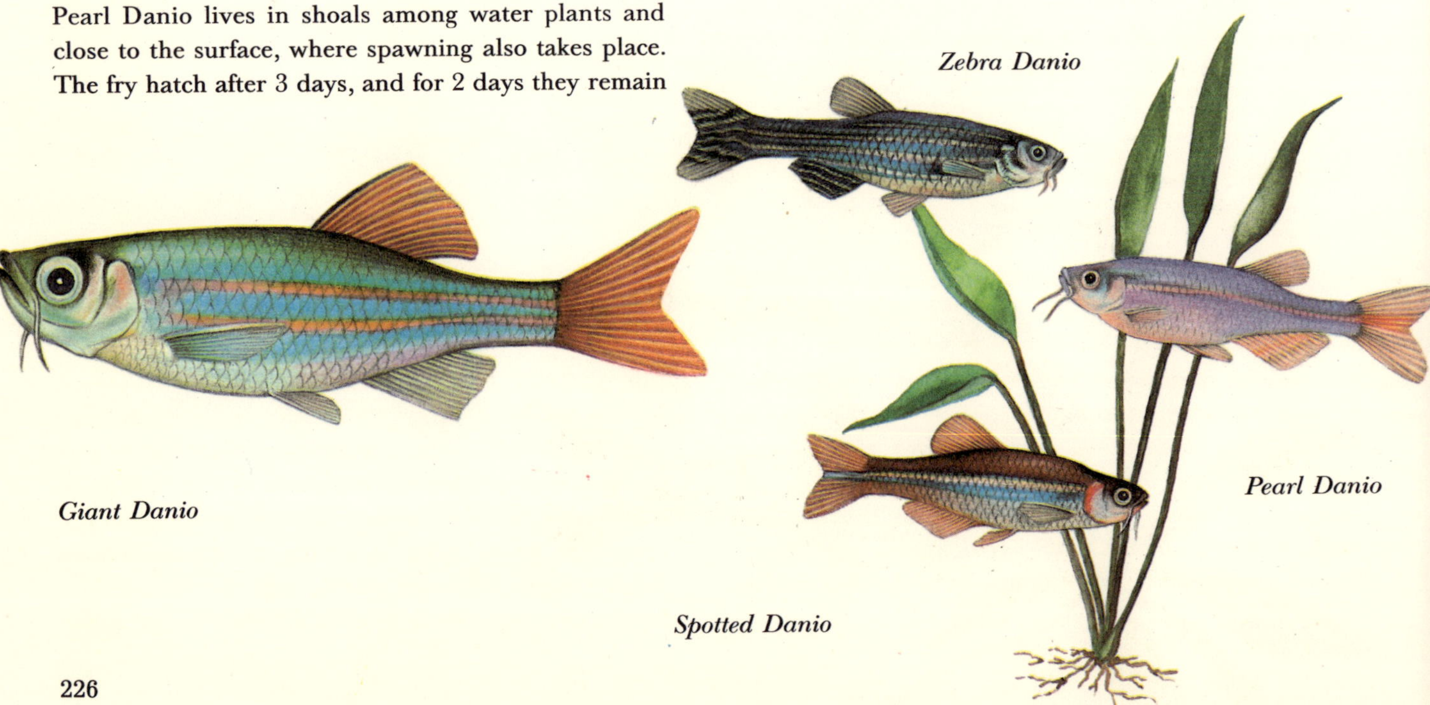

Zebra Danio

Pearl Danio

Giant Danio

Spotted Danio

included in its diet. Although this species is very
abundant, little is known of its reproductive behaviour.

COOLIE LOACH
Acanthopthalmus kuhli kuhli

The Coolie Loach is about 10 cm long and is found in
the waters of Sumatra and Java. It frequents still waters with muddy or sandy bottoms densely covered
with weeds. It also occurs in lowland and mountain
streams. During the day, this fish lies buried in the
mud, coming out to hunt only after nightfall. It feeds
on tiny creatures, seeking them on the bottom with
three pairs of touch-sensitive barbels situated around
its mouth. It also eats plant material. When alarmed,
it starts up sharply, but immediately settles down
again a little further away. At spawning time, the
fishes swim up to the surface in a spiral movement.
The female may lay several thousand eggs.

CLOWN LOACH
Botia macracanthus

The Clown Loach, which is about 30 cm long, frequents rivers and large streams in Sumatra and
Borneo. It is an attractively coloured fish, often bred
in aquariums, where it reaches a length of only 15 cm.
It lives in shoals along overgrown banks. Although it
is omnivorous, it prefers animal food and preys mainly upon worms and insect larvae.

GLASS CATFISH
Kryptopterus bicirrhis

The Glass Catfish is distributed from India across
south-eastern Asia to the Sunda Islands. It reaches
a length of 10—11 cm, and has a vertically flattened
body. The dorsal fin is formed by a single ray, while
the anal fin is considerably elongated. There is a pair
of long barbels on the upper jaw, and the body is so
opaque that the outlines of the organs can be seen
against the light. This species is found in rivers and
streams in low-lying regions, living in small shoals
close to the bottom. Crustaceans, worms and insect
larvae, especially those of gnats, form the diet of this
fish.

Red-tailed Black 'Shark'

Coolie Loach

Clown Loach

Glass Catfish

MALABAR KILLY
Aplocheilus lineatus

The Malabar Killy, a beautifully coloured species,
lives in low-lying waters of southern India. Its average
length is about 10 cm and its coloration varies with
the area of distribution. The female is less striking in
colour and the protuberance on her caudal fin is not
opaque as in the male. Spawning takes place among
dense tangles of water plants, and the female lays
large sticky eggs which become attached to the aquat-

Malabar Killy

Indian Glassfish

Orange Chromide

Badis

Siamese Fighting Fish

ic vegetation. The fry hatch out after 14—17 days and feed on insect larvae and worms. The adults also prey on tiny fishes.

INDIAN GLASSFISH
Chanda ranga

The Indian Glassfish is found in India, Burma and Thailand. It attains a length of 7 cm in the wild, but when kept in an aquarium, it is always much smaller. It has an opaque body. This fish frequents lowland rivers and is often found in the brackish waters of estuaries. In the spawning period the female lays her eggs on water plants. The young hatch after 24 hours and begin to swim when they are 3—4 days old. At first, they move with their heads turned upwards, and in this vertical position they catch minute crustaceans. After a few days, they begin to swim like other fishes. This species feeds entirely on live prey, such as insect larvae, crustaceans, worms and molluscs.

ORANGE CHROMIDE
Etroplus maculatus

The Orange Chromide reaches a length of up to 8 cm and has a very deep body. It is a denizen of both fresh and brackish water in India, and prefers localities where the vegetation is thick. The female lays her eggs on stones or roots along the banks, and here the fry remain attached for 2—3 days before dispersing. At first they feed on tiny live prey, later concentrating on crustaceans. Adult fishes feed predominantly on vegetable material, though crustaceans, insect larvae and worms form a part of their diet.

BADIS
Badis badis

The Badis lives in standing water throughout India. The male reaches a length of 8 cm, while the female is about 5 cm long and is predominantly reddish-brown in colour. Before spawning, the male carefully cleans the spot chosen for egg-laying, and the female lays her eggs in crevices among the stones. The eggs and the fry, numbering about 40—60 tiny fishes, are guarded by the male. This fish feeds on various small creatures, caught mainly at dusk.

SIAMESE FIGHTING FISH
Betta splendens

The Siamese Fighting Fish is one of the best-known aquarium fishes. The veiled forms of various colours — blue, green, red, yellow, black — originated in captivity. This fish inhabits still or sluggish waters of the rivers of the Malay Peninsula and Thailand. In the spawning period, the male builds a spongy bubble-nest which floats among weeds near the surface. After the female has laid her eggs in the nest, the male chases her away and guards the clutch on his own. The fry hatch 24—30 hours after spawning and are watched over by the male for the first few days of life. Young fighting fishes mature at the age of 4 months. This species feeds on tiny invertebrates and pieces of plant material.

Male fighting fishes are very pugnacious. At one time it was customary in Thailand to put rival fishes in glass bowls and to bet on the results. In the courtship season, the males become particularly aggressive and engage in fierce duels.

DWARF GOURAMI
Colisa lalia

The Dwarf Gourami is one of world's most beautiful fishes. In contrast to the male, the female is predominantly brownish with green stripes. This fish, which reaches a length of about 5 cm, inhabits three Indian rivers, the Ganges, the Yamuna and the Brahmaputra and its tributaries. It lives near the banks of coves, swimming among rooted or floating weeds. The eggs are embedded in a spongy nest of bubbles, built by the male near to the surface of the water, and it is the male which ensures the safety of both spawn and fry. The Dwarf Gourami eats tiny invertebrates and pieces of plant material.

STRIPED GOURAMI
Colisa fasciata

The Striped Gourami, which is about 12 cm long, is found in the deltas of the Indian rivers Ganges and Brahmaputra. It also lives in the streams of north-west India and Burma. In the nuptial period, the male becomes exceedingly aggressive. Breeding behaviour is similar to that of the Dwarf Gourami.

HONEY GOURAMI
Colisa chuna

The Honey Gourami is a tiny fish, about 4 cm long. The female is greyish in colour. This species lives in coastal rivers of India, particularly in the Brahmaputra, frequenting sites overgrown with vegetation. The male builds a small spongy nest of bubbles, able to contain only a few eggs. Breeding and diet are as for the Dwarf Gourami.

MOSAIC GOURAMI
Trichogaster leeri

The Mosaic Gourami is native to Malaysia, Thailand, Borneo and Sumatra. It measures 11—12 cm in length and the female is less colourful than the male. This species lives in schools which inhabit streams and flooded areas of rivers flowing through both lowland and hilly localities. The male builds a foam nest into which the female lays as many as 2 000 eggs, and the nest and fry are guarded by the male. The food of this fish is small invertebrates and plant material.

THREE-SPOT GOURAMI
Trichogaster trichopterus trichopterus

The Three-spot Gourami, which attains a length of 15 cm, inhabits the waters of the Greater Sunda Islands, Malaysia, Thailand and southern Vietnam. It is common in low-lying areas, but also occurs in mountainous districts. There are several other subspecies differing in coloration. The Blue Gourami (Trichogaster trichopterus sumatranus) is azure-blue in colour. Its breeding and diet are similar to those of the previous species.

PARADISEFISH
Macropodus opercularis

The Paradisefish Macropodus opercularis is indigenous to China, Korea, Taiwan and southern Vietnam. It measures 9—10 cm in length. The female has shorter fins than the male and is more subdued in colour. Paradisefishes were among the first aquarium species to be bred in captivity. In the wild they inhabit muddy, often shallow water, and they are sometimes even

Mosaic Gourami

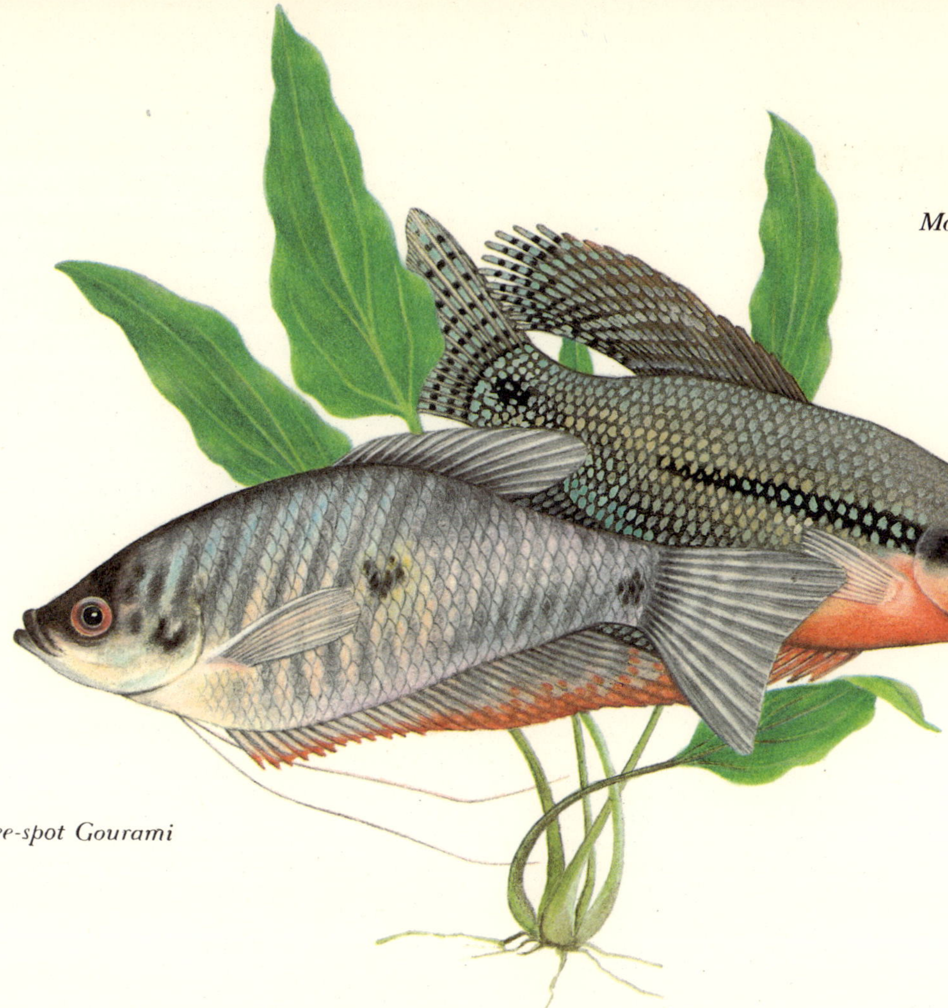

Three-spot Gourami

found in flooded rice paddies. During spawning, the male approaches his mate, spreads his fins and becomes darker in colour. The female goes to meet him, often swimming in circles in a vertical position, and the two fishes sometimes chase one another. The male builds a foam nest for the eggs by swimming up to the surface, swallowing air and discharging saliva in the form of bubbles under the water. The bubbles rise and form a nest-like shape near the surface. The female lays her eggs in the foam and the fry hatch out 36 hours after spawning. The male guards the young and keeps them in a group, but after a few days he takes no further care of them.

Paradisefishes feed on insects, larvae, molluscs and worms, and they sometimes prey on the eggs and fry of other species of fish.

Paradisefish Macropodus opercularis

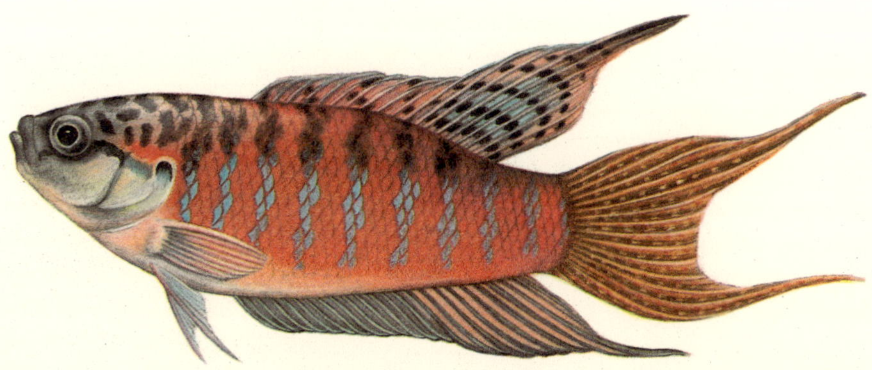

Brown Spike-tailed Paradisefish

BROWN SPIKE-TAILED PARADISEFISH
Macropodus cupanus dayi

The Brown Spike-tailed Paradisefish lives in the waters of southern India, Burma and southern Vietnam. It is about 7.5 cm long. In the spawning period, the male's abdomen becomes a luminescent red colour. Breeding behaviour and diet are similar to those of the previous species.

GREEN PUFFER-FISH
Tetraodon fluviatilis

The Green Puffer-fish is a bizarre species about 20 cm long, which is distributed from India to the Philip-

pines. It inhabits rivers and the brackish waters of estuaries. It is a slow swimmer but can surface speedily or drop swiftly to the bottom to escape its predators. This fish has an intestinal sac which can be quickly filled with air. This enables the puffer-fish to swell up like a balloon, and so deter its predators. It feeds mostly on molluscs, easily crushing their shells with its four powerful teeth. During spawning the male holds the female firmly with his teeth.

PUFFER-FISH
Tetraodon palembangensis

The puffer-fish *Tetraodon palembangensis* can grow to a length of 20 cm. It is very common in fresh water on the islands of Borneo and Sumatra and in Thailand. Its breeding behaviour and diet are similar to those of the Green Puffer-fish.

Green Puffer-fish

Puffer-fish Tetraodon palembangensis

COASTAL
AREAS
OF TROPICAL
ASIA

———

The shores of tropical Asia are bounded by the Indian Ocean in the south, and by the Pacific Ocean in the east. The coastline, however, is washed by the waves of many seas, including the Arabian Sea, the Andaman Sea, the South China Sea, the Java Sea and the Celebes Sea. Sandy lowlands characterize the local landscape, with steep rocks and low cliffs emerging in some localities. Extensive sandy beaches cover large areas. Massive river estuaries are a particular feature of the coastal regions, with their typical cover of mangroves. The genera most frequently found are *Rhizophora, Avicennia, Sonneratia, Ceriops* and *Carapa.* Dense, impenetrable mangrove swamps, often many kilometres long and several hundred metres wide, cover the flat, muddy shores, quiet bays and lagoons where the water is calm. The evergreen, multibranched mangroves provide shelter for many wading birds such as herons and marabous, and for raptors which find easy pickings near the colonies of fish-eating birds. Mangroves are adapted to the changes in water level as the tide ebbs and flows. They have still roots which support the trunks above the water and there are also aerial roots which ensure an adequate supply of oxygen to the root systems.

Along the coasts and on the islands off the coasts of India and Indonesia, there are vast forests of palms of the genus *Nypa,* which grow to a height of 4—6 m. In estuaries, these palms are found at the limit of the tidal flow, but they always grow in brackish water, and often merge with belts of mangroves.

People travel considerable distances on foot or by boat, to collect the huge leaves of these palms, cutting them off with machetes. The leaves are used to roof the wooden huts. Juice from the flowers of *Nypa fruticans* is fermented to make an excellent wine.

Screw pines of the genus *Pandanus* have dense, tough sword-like leaves with spiky, saw-like edges. These palm-like trees have prop roots, and their crowns resemble gigantic helmet crests. There are a great many species in the Asian tropics, reaching 3—6 m in height. The Malaysian coast is covered by extensive screw pine forests, but they also occur inland, on the edges of swamp pools and river banks. Their leaves yield fibres used in the weaving of mats, sunhats, sandals, and even sails. The pulpy fruit is edible, although it is not especially popular.

Of all the trees of the tropical coasts of Asia, the most important is the well-known Coconut Palm *(Cocos nucifera)*. This tree is native to the coral islands and shores of Polynesia, but it has been widespread throughout the Indo-Australian regions since ancient times. In many areas, it is a semiwild plant, while in other localities it is cultivated in various forms in huge plantations. Coconut palms grow only in humid coastal regions reached by winds blowing from the sea, and the trunks are often bent or distorted. They are very common along the coasts of Sri Lanka and south-east Asia and Java, where they are of great significance in the nutrition of thousands of the inhabitants. The trunks of these palms tower to a height of over 20 m and are surmounted by crowns of arc-shaped leaves, up to 5 m long. Among the leaves the coconuts grow in clusters. The nuts are about 30 cm long and are among the biggest fruits in the plant realm. Each coconut is encased in a thick husk of tough, brown fibres. Underneath is the large nut with a hard but thin shell surrounding the kernel. The whitish, oily kernel is hollow, and before ripening is filled with a sweetish, watery liquid, called coconut milk. When the coconuts are ripe, they fall off and usually remain on the ground near the palm. Sometimes, however, they may be seized by a wave or they may drop directly into the water and be carried away by the current. Their hollow structure enables them to float and they often reach localities hundreds of kilometres away, where the waves throw them on to the beach. Here they sprout and eventually give rise to a new colony. Coconut palms flower and produce fruit after only 6 years, and a single palm can yield 50—100 coconuts annually. About 6 000 coconuts are harvested every year from each hectare of plantation. Coconut palms are put to a variety of uses. The wood is used to make furniture, the huge leaves are made into matting and roofing, and the juice is extracted from the flower buds, and fermented to make palm wine, or boiled and thickened for its brownish sugar. Of greatest value is the fruit, which is picked by men who climb the sloping trunks using only their arms and legs—and a piece of rope as well if they climb up a vertical palm trunk. They chop off the ripe coconuts with machetes.

Attempts have been made to train monkeys to pick the coconuts and throw them down, but this has not proved successful. Coconuts growing on smaller palms are harvested with hooks attached to long bamboo sticks. Everywhere along the coast, the markets are full of coconuts. A freshly picked nut contains up to a pint of coconut milk, and makes a refreshing drink in the hot tropical climate as well as being nutritious. The major product obtained from coconuts is the dried flesh, called copra. When pressed the copra yields large quantities of vegetable oil, the residue being fed to cattle. Shredded copra is used in many forms of cooking all over the world, and as an ingredient in confectionery. The fibrous husk is processed to make a material called coir which is used in the manufacture of rugs, mats, coarse cloth, brushes and padding. Coconut fibres are very light and water-resistant. The shells are made into receptacles, spoons and many other objects. It is interesting to note that the first historical references to coconut palms date from an expedition to Sri Lanka by Alexander the Great, in the 4th century B.C.

Another tall palm, the Date Palm *(Phoenix sylvestris)* is found in India. The sugary sap from its trunks yields syrup which is used to make tari, a kind of palm wine.

Round all the coasts of Asia beautiful yellow-flowered opuntias *(Opuntia)* grow to a height of 4 m. In many places, these massive cacti are used instead of fences to protect gardens and cultivated crops, for the spines on the dense branches deter even large animals. Opuntias are not native to Asia but originated in the American tropics. Their fruit is edible. Sparse woods of pine trees can also be seen along the coasts, although the trees are only a few metres high, and small groves of evergreen plants of the genus *Ephedra* are common in sandy localities.

When the tide ebbs over the extensive beaches of tropical Asia, it reveals sand teeming with all kinds of creatures, expecially crabs of many species and all sizes. Thousands of these crustaceans leave the water and run sideways on the wet sand, washed by the foaming surf. The crabs dig deep tunnels in the sand, where they can hide when danger threatens. For most of the time, they stay near these tunnels, searching for scraps of vegetable or animal material on which to feed. They scoop up their food with their claws, along with a quantity of sand, and pass the mixture to their mouths. Then they separate the edible parts and discard the sand grains and the unwanted parts of the food, moulding it into balls. The balls can be seen around the entrances to the underground holes, until the next tide carries them away. Local crab-hunters

find the biggest crabs by searching for the largest entrances. Then they dig out the crabs and put them in bamboo baskets. They cook them at home or sell them in the markets, for the fine crab meat is regarded as a delicacy. The crabs vanish when the tide comes in, but reappear when the water ebbs away again. Crabs are also preyed upon by a variety of wading birds as well as by kites and crows.

The receding tide also exposes many other sea creatures, thrown on to the beach by the waves and unable to get back to the sea. These animals soon perish in the scorching sun. Among them are many jellyfishes, the gelatinous bodies of which soon turn into shapeless masses on the sand. It is unwise to touch them, even in this state, for the stinging cells on their tentacles can inflict long-lasting, burning wounds. Sea snakes, although they are venomous, are helpless when they are out of their element and soon fall prey to kites, crows and small carnivores. The tide also brings in various molluscs, especially sluggish bivalves, which have shells consisting of two hinged parts. Clams and mussels are a welcome source of food for the local people, who catch them when the tide is ebbing. Women use baskets with sieve-like bottoms to scoop up the water, and when the fine grains of sand are washed away, the tiny shellfishes are left behind. Thousands of minute bivalves are eaten every day. The women boil them in water, pick out the flesh and then discard the shells. The villages are often surrounded by heaps of shells glittering like jewels in the sunshine. The men hunt larger bivalves which are hidden in the sand in deeper water further from the shore. They wade up to their waists in water dragging behind them bamboo sticks fitted with hooks. The hooks pass easily through the fine sand, but they catch on large shells. The hunter digs out the mollusc and puts it into a bag made of coconut fibres before continuing his search.

Fish is the main food of the coastal populations. Fishermen set out every day in wooden boats with paddles and sails, in small barges, or on rafts. In some regions, rafts are made from long bamboo stalks, tied together with lianas. These peculiar craft have a bamboo mast in the centre. It is remarkable that pairs of fishermen venture out to sea in these primitive vessels, but they set off in the morning when the tide ebbs, and return at high tide in the late afternoon. In the warm seas, their catches often include various species of huge sharks. The Six-gilled Shark *(Hexanchus griseus)* measures up to 3 m, the Hammerhead *(Sphyrna zygaena)* grows to a length of 5 m, and the dangerous man-eater, the Great White Shark *(Carcharodon carcharis)*, can reach a length of 12 m. Small octopuses often become entangled in the nets and end up on the table as a meal. Some fishermen use large basket-like traps, about 2 m high, 3 m long and 2 m wide, made of splintered bamboo stalks, submerging them in the water and returning the next day to take out the fishes. The bamboo cage has a long, tunnel-shaped entrance through which the fishes can easily swim to get inside, but they are prevented from getting out by sharp spikes.

The area of the Ganges delta provides ideal habitats for tigers, which prowl the mangrove jungles and vast reed beds. Man-eating tigers are more common here than in other areas, for there may be as many as 10 tigers in every square kilometre, and when supplies of their normal food are low, hunger forces these carnivores to kill people. According to the records, tigers accounted for the deaths of 126 men in the three years from 1969—1971. In the delta, such animals as hogs and various mammals have been extensively hunted by the local population with the result that the people themselves now run a greater risk of being attacked by hungry tigers. However, it is not only the tigers' food sources that are diminishing. The number of tigers is also falling, although they used to be common. In many places, they have been completely exterminated. Only thirty years ago, over 50 000 Bengal Tigers *(Panthera tigris tigris)* lived in India, but now there are only about 2 000. Other subspecies have suffered even more.

In the early 20th century, the entire continent of Asia was inhabited by something in the region of 100 000 tigers. Today, zoologists estimate that Asiatic tigers number about 4 000. The original eight subspecies have been reduced to six, and even these are on the verge of extinction. The smallest subspecies, the Bali Tiger *(Panthera tigris balica)*, has been exterminated without any hope for the future. The Caspian Tiger *(Panthera tigris virgata)*, which lived in the area around the Caspian Sea and in the territory stretching from Afghanistan to eastern Turkey, is probably extinct. Some experts believe it could still be found in some remote regions, but this is probably wishful thinking. The Javan Tiger *(Panthera tigris sondaica)*, with just 5

specimens left on Java, faces extinction. The Sumatran Tiger *(Panthera tigris sumatrae)* survives in the wild on Sumatra, with about 400 specimens. There are only a few specimens of the South China Tiger *(Panthera tigris amoyensis)* left. The Ussurian Tiger *(Panthera tigris altaica)*, also called Siberian Tiger, has been reduced to about 250 animals living along the Amur, and there are 2 000 Indochina Tigers *(Panthera tigris corbetti)* in the area of Indochina.

Another large Asian carnivore is the Indian Lion *(Panthera leo goojratensis)*. This species is now restricted to a small area on the Kathiawar Peninsula, near the coast of north-western India. This rare animal is strictly protected in the steppes and woodlands of the Gir Wildlife Sanctuary. The landscape here is open, being predominantly covered by scrub, tall grass and scattered teak trees, and the sanctuary contains a number of villages. The lions hunt mainly for wild hogs, antelopes, axis deer and sambars. At one time lions inhabited vast areas of western, northern and central India, but they had been exterminated in most of these places by 1880, only a small population remaining in Kathiawar. The numbers, unfortunately, are still falling, despite the fact that lion hunting is strictly prohibited, and only a few animals are made available to zoos and wildlife sanctuaries, for breeding purposes.

WHITE-BREASTED SEA EAGLE
Haliaeetus leucogaster

The White-breasted Sea Eagle is found mainly in coastal areas, but it also inhabits large lakes and estuaries ranging from India across south-eastern Asia, the Greater Sunda Islands and the Philippines to Australia. The male reaches a length of 70 cm and has a wingspan of 220 cm, the female being somewhat larger. Adult sea eagles form pairs for life, while juvenile birds roam singly. Nesting takes place between September and January. Each pair builds a large nest of branches and sticks in its extensive territory. The nest is about 150 cm across and 75 cm high, and is usually situated 10—50 m above the ground in a gigantic casuarina tree growing on the coast. Sea eagles sometimes build their eyries on rocky outcrops on off-shore islands. They use the same nest for several years, one having been seen to return to the same nest for 50 years. The birds repair their nest regularly and continually enlarge it. The female lays 2 white eggs, which are incubated by both partners for 35—40 days. The young are also fed by both their parents. After 65—75 days the young eagles leave the nest and learn to fly and hunt. The White-breasted Sea Eagle feeds on sea snakes, caught in shallow water near the shore or seized as they are thrown up onto the beach by the waves. They also catch fishes on the surface by dipping their feet into the water as they fly and piercing the prey with sharp talons. Crabs are also eaten, and sea eagles sometimes take the catch in fishing harbours. These eagles often hover in one spot about 15 m above the water, scanning the surface for prey.

YELLOW-LEGGED BUTTONQUAIL
Turnix tanki

The Yellow-legged Buttonquail inhabits coastal scrubland, grassland and fields from India to south-eastern

White-breasted Sea Eagle

and eastern Asia. It reaches a length of 17 cm and the female is much more impressive in appearance than the male. This buttonquail lives in thick vegetation where it feeds on seeds and tiny insects, and pecks green plants. The nest consists of a shallow scrape in the grass. Here the female lays 4 spotted eggs, which are incubated by the male. The female takes no interest in the eggs or in the brood, but joins another male. The young hatch out after 15 days and their father guides and protects them.

GREAT STONE CURLEW
Esacus recurvirostris

The Great Stone Curlew is found along the seashores and the banks of large rivers in an area stretching from Palestine to south-eastern Asia, the Greater Sunda Islands, the Philippines and Australia. This bird reaches a length of 57 cm and is characterized by a huge, slightly hooked bill. Its favourite habitats are sandy or swampy localities near the coast or around lagoons, far from human dwellings and where the birds are not disturbed. The Great Stone Curlew probably lives in pairs throughout its life, but outside the nesting season the pairs gather in small groups of up to 20. This bird is predominantly nocturnal, foraging after nightfall. During the day it rests on the banks, often in the safety of an island. When danger threatens, the curlew warns other birds by means of sharp cries and flies away. This species preys chiefly upon crabs, breaking the carapaces with its powerful bill, or smashing larger specimens against a stone. It also feeds on molluscs, worms, insects and other invertebrates, and sometimes takes the eggs of other birds.

Nesting takes place between January and August. The female lays 1—2 eggs, greyish-yellow in colour, and densely streaked with brown and black spots. The nest is an unlined shallow scrape on the bank of a stone-covered island or on a sandy shore. Incubation is carried out by both partners for 26 days, and the brood is reared by both parents. The bird which is not sitting stays near at hand to warn its mate of any danger. The sitting bird runs a few metres away from the nest before taking to the air. On its return, it is very cautious, approaching one step at a time, and looking carefully around as it walks. At night, young curlews follow their parents to search for food.

CRAB PLOVER
Dromas ardeola

The Crab Plover is an extraordinary species of plover. It is about 40 cm long, and has an unusually large head, long legs, and a massive bill. It is found on the coasts of Sri Lanka, western India, the Andaman and Nicobar Islands, and on the eastern shores of Africa. Outside the nesting season, it also flies to the coasts of Malaysia. It lives on remote sandbanks and rock-

Great Stone Curlew

Crab Plover

covered areas, usually in pairs or small groups, although solitary plovers can also be seen, and sometimes over 100 plovers rest together on the cliffs. These birds sing melodiously as they fly in tight formation. Crab plovers forage in the tidal zone or on sandbanks, where they find thousands of crabs, carried in and left by the tide. They kill the crabs with their beaks, break the strong carapaces and peck out soft pieces of flesh. Small crabs they swallow whole. Crab plovers also feed on marine molluscs and worms, and often hunt in the low water of lagoons.

In Asia, nesting takes place in May and June. The plovers excavate curved passages 60—120 cm long in sandbanks, or use the burrows of large crabs. The tunnel ends in a nesting chamber, where the light cannot reach and here the female lays a single white egg. Both partners take part in incubation for 24 days, and they both feed the young chick on crab meat. The chick can run on the second day after hatching, but it usually stays in the burrow for about 14 days. When it leaves the nest, it runs about with its parents, but it always returns to the nest when it is in danger.

Common or Collared Pratincole

COMMON or COLLARED PRATINCOLE
Glareola pratincola

The Common or Collared Pratincole is widespread in several subspecies from eastern Europe across central Asia to eastern Asia, Sri Lanka, and throughout the Greater Sunda Islands and the Philippines to Australia and New Zealand. It attains a length of 25 cm and both sexes are identically coloured. This bird inhabits open country, especially steppes, along the coast of the Caspian and Black Seas, but it also occurs in brackish marshes situated inland and near water. It lives in groups and often nests in colonies, the nests being usually 50—100 m apart. They are only shallow depressions, mostly unlined, but a pair often uses the same nesting spot for many seasons. The female lays 2—3 eggs, greyish-yellow with brown and grey spots, at some time between March and June, and the clutch is incubated by both partners for 18 days. The young birds hide in tufts of grass, among boulders or bushes, and the adults bring food for them. The Common Pratincole preys in flight on butterflies, beetles, winged termites and various other winged insects. It sometimes also catches insects on the ground.

INDIAN COURSER
Cursorius coromandelicus

The Indian Courser inhabits India and northern Sri Lanka. It is about 26 cm long and has a short tail and extremely long limbs, adapted for fast running. The courser scurries over the ground, taking short steps, stopping abruptly to pick up an insect, and running on. It is a good flier as well. Coursers are found only on sandy or stony terrain where there is a thin cover of grass. Here they mainly catch locusts, grasshoppers, beetles and termites. Outside the nesting season, they wander in small flocks or in pairs. In May or June the female lays 2 eggs in an unlined depression on the ground. The eggs are yellowish, and are covered with purplish-grey and reddish spots. The female incubates the eggs for 18 days, while the male stays on guard. He helps to rear the young, which often leave the nest on the second day after hatching.

SLENDER-BILLED GULL
Larus genei

The Slender-billed Gull is found in coastal areas of Pakistan, the Persian Gulf and the Caspian and Black Seas. Its distribution is discontinuous. This gull is about 43 cm long. It nests in colonies, usually on rocky or sandy islands, each colony comprising up to 100 pairs with nests about 70 cm apart. In May or June, the female lays 2—3 eggs in a nest made of water plants and grass. The eggs are variable in colour, usually being yellowish-white with black or brown dots, and grey or violet patches. Incubation is shared by both partners for 26 days. Two or three days after hatching, the young, which are covered with a dense layer of down, leave the nest and hide in tussocks of grass or among stones, where their parents bring them food. The Slender-billed Gull feeds on small fishes, insects, crustaceans and molluscs.

BROWN HAWK OWL
Ninox scutulata

The Brown Hawk Owl is widely distributed in several subspecies in southern, south-eastern and eastern Asia, in the Sunda Islands, the Philippines and Sulawesi. It is about 32 cm long. It inhabits forests bordering rivers and stretches of water near the coast, but it also occurs inland. In mountainous areas it lives at a height of up to 2 000 m. During the day the owl sleeps in a regular site among dense branches. Pairs often stay together and the partners sit next to each other. The Brown Hawk Owl hunts after nigtfall, but on overcast days it sometimes begins during the day, mainly in the morning. It is an expert flier, catching small birds, bats, winged termites, butterflies and beetles on the wing. It also snatches small lizards, frogs, and occasionally even snakes.

Indian Courser

In the nesting season, which takes place from January to March, the owl makes deep croaking sounds, repeated six to twenty times. The nest is constructed in a hole in a large tree, usually a mango, 2—5 m above the ground. The female lays 2—3 white eggs and sits on theem for 24 days, being relieved at intervals by her mate.

LEATHERY TURTLE
Dermochelys coriacea

The Leathery Turtle is distributed in warm tropical seas. It reaches a length of 3 m and a weight of

Slender-billed Gull

Brown Hawk Owl

on the eastern coast of Malaysia, in the state of Trenggan, where the leathery turtles lay their eggs in May and June. The female crawls laboriously to a point above the line of the tide where she excavates a hole about 60 cm deep, and within an hour she has laid the whole clutch. She then buries the eggs in the sand, treads the ground down around them, and returns to the sea. The young hatch after 2 months and immediately hurry to the water. At this time many enemies gather on the shore, particularly birds of prey and carnivores, and sharks wait in the water. However, man is this turtle's biggest enemy, for the eggs are collected on a large scale and eaten. The species was being threatened by imminent extinction when the zoological department of the Malaysian University first made a move to save the leathery turtles. Collectors are now paid to find the eggs. Clutches are easily found because the females leave wide tracks in the sand. The collectors gather thousands of eggs in the threatened localities and bury them in a fenced-in site on the coast near Dungun. Up to 10 000 eggs are guarded here, and the clutches are marked. After the eggs have hatched the university workers remove the tiny turtles and release them in the sea. Every year several thousand leathery turtles are saved in this way.

1 tonne. It has a flat body and a compressed, leathery carapace, with no horny plates. The carapace is small and irregular. This turtle can remain underwater for as long as 50 minutes. When sleeping, it has to swim periodically up to the surface to breathe. The male never leaves the sea, but in the egg-laying season the female swims ashore to lay her clutch of 90 or so eggs. Only a very few nesting beaches are known, one being

HAWKSBILL
Eretmochelys imbricata

In Asia, the Hawksbill inhabits the warm Pacific and Indian Oceans. Its carapace is up to 90 cm long, and the species is recognized by the overlapping, tile-like arrangement of the plates of the carapace. The Hawksbill obtains its name from the drawn-out upper

Leathery Turtle

Hawksbill

jaw which resembles a hooked beak. This turtle roams in the seas, preying on various small marine animals and eating marine plants. The females come ashore to their specific nesting grounds, usually at night. They dig out holes about 50 cm deep with their hind limbs, and each female lays 100—150 eggs. Then they cover the clutches with sand and return to the sea. The young hatch out in about 50—60 days, and immediately seek refuge in the water. Before reaching it, many are eaten by rapacious birds and animals who wait for days for the tasty, helpless prey.

The Hawksbill is one of the most sought-after turtles, on account of its beautiful, high-quality carapace. At present, this species is internationally protected in the same way as other marine turtles. Nevertheless, it is still hunted for its tortoiseshell, and local people also eat the meat.

SALT-WATER CROCODILE
Crocodylus porosus

The Salt-water Crocodile has a vast area of distribution, covering southern India, Sri Lanka, Sumatra, Java, Borneo, Sulawesi, the Philippines, New Guinea and the northern coast of Australia. It can reach a length of 7 m, and specimens about 5—6 m long are common. This crocodile lives in coastal seas and in the brackish waters of estuaries. It often swims far out to sea and settles on islands where it has never occurred before. Some crocodiles have reached localities over 1 000 km from their original habitat.

The Salt-water Crocodile is one of the few species of crocodile that does not hesitate to attack man. On some popular beaches there are special watch-towers from which patrols scan the sea with binoculars, in

Salt-water Crocodile

case a stray specimen appears. A horn is sounded to warn the bathers who hurry out of the water while the guard fires to deter this dangerous reptile.

Crocodiles build their nests on sandy banks. The nests consist of a large heap of mud or sand, together with leaves, sticks and other pieces of plant material. The female lays her eggs, covers them with soil, and stays on guard nearby. She also watches over her young for a few days after hatching. Baby crocodiles are preyed upon by monitors and wild hogs, and also by male crocodiles. In the mating season, the male makes a call resembling the lowing of a cow.

The Salt-water Crocodile feeds on fishes, mammals and birds, the size depending on the size of the crocodile. The young also take insects and other invertebrates.

BELL'S AGAMA
Leiolepis belliana

Bell's Agama is found along the coasts of south-east Asia, and can exceed 40 cm in length. While the male is beautifully coloured, the female has inconspicuous olive-brown coloration with dark dots. This agama lives on the ground in sandy regions near the line of the tide. Here it digs holes up to 50 cm deep, in which to shelter at night and during the midday heat. In the early morning and in the afternoon, when the sun's rays are less scorching, the agama begins to forage. It hunts insects, spiders, and small crustaceans, especially crabs, and it also eats vegetable material. The female lays several eggs with membranous shells in holes dug out in the sand. The young hatch in about a month, then disperse to seek their own territories. In some areas, fishermen dig out the agamas from their burrows and use them as bait.

KOMODO DRAGON
Varanus komodoensis

The Komodo Dragon is the largest, and probably the best-known of the lizards. It inhabits Komodo Island, the western part of the island of Flores, and the small island of Rintja. It can reach a length of 3 m — in fact the largest reported specimen measured 3.1 m. It weighs up to 135 kg.

The Komodo Dragon was only discovered in 1912, and its discovery caused a sensation not only among laymen, but also among scientists. Although the inhabitants of adjacent islands claimed that there were 'dragons' on Komodo, nobody believed them, and certainly scientists paid no attention to such reports. But one day, a pilot was forced to crash-land his aircraft on the island. Though he had heard about the dragons, he did not believe the stories either. However, when searching for water on the island, he found a strange, wide track on the sandy shore, and then he saw the monsters. He had to spend several months on the island before he was rescued, and when he returned home, nobody accepted his story about the dragons. Everyone thought that living alone on the island had driven him mad. Some time later, a Dutch hunter visited the island and caught one of the dragons, which was subsequently scientifically described. The island then became the focus of many expeditions. The natives asserted that the animals were about 10 metres long and killed and devoured men, but no scientific expedition succeeded in capturing a dragon longer than three metres. The island was soon besieged by hunters, all eager to capture a dragon, and soon the rare lizards were in danger of extinction. At this point the island of Komodo was declared a reserve.

The Komodo Dragon feeds on wild boars and small

Bell's Agama

Komodo Dragon

species of deer. Young dragons climb up trees and plunder birds' nests, eating both eggs and nestlings. They also catch adult pigeons and parrots. On the island of Rintja, which is inhabited by monkeys, they catch these as well. When a dragon dies the other dragons immediately tear up its body and devour it. About 150 natives live along the coasts of Komodo, and they maintain that the dragons also preyed on men. However, there are only two confirmed cases of Komodo dragons attacking men. In one case a dragon pursued three little boys and killed one of them, and in another a dragon attacked two men who chased it away. There have been no other similar cases for decades. Komodo dragons shelter in burrows which they dig in relatively inaccessible places. In summer, the female lays 5—20 eggs 10 cm long and 6 cm wide, and buries them in the sand.

EARLESS MONITOR
Lanthanotus borneensis

The Earless Monitor is a bizarre primitive lizard, taken by some scientists to be the ancestor of the snakes. It has very short, underdeveloped legs and an elongated body, which reaches a length of about 40 cm. It has a large head, and no external ears. This rare, nocturnal lizard was first discovered in Sarawak, in the north of the island of Borneo, in 1878. Only ten other

specimens have been captured since, all in the same area, but in 1961 one of them survived in captivity after being transported from Borneo. The places where this monitor is to be found are virtually inaccessible to man. The species lives in holes, tree hollows and underneath dense bushes. It is able to burrow in moist soil, and it is a good swimmer. It moves in a rather snake-like manner, pushing with its feet.

RACERUNNER
Eremias arguta

The racerunner *Eremias arguta* is widespread, being found in several subspecies from north-eastern Romania across northern Iran to central Asia and Trans-Caucasia. It lives in a variety of habitats, being often encountered along the sandy shores of the Black Sea and the Caspian Sea. It is also found on steppes, along the edges of forests, and in mountains at a height of 2 100 m. This lizard reaches a length of 15 cm. It digs holes about 5—10 cm deep in sand or soft soil, or hides beneath stones, often in company with scorpions. It hunts in the morning or in the afternoon, scurrying in the neighbourhood of its burrow mainly catching beetles, although it occasionally eats locusts, various other insects, and spiders. In late May or early June, the female lays 4—12 eggs, which usually hatch in July.

ELEPHANT'S TRUNK SNAKE
Acrochordus javanicus

The Elephant's Trunk Snake is very common in Malaysia, Thailand, the Sunda Islands and New Guinea. It reaches a length of 2.5 m, and has a short, broad head. It spends all its life in water, where it preys on fishes and frogs. It seldom goes ashore, and it moves clumsily on land. This snake abounds in river deltas, where fishermen often catch it in their nets. Although the species is not venomous, the fishermen have to take care to avoid the snake's teeth, for they can inflict deep wounds. The fishermen take the snakes home in baskets and skin them, for these snakes have valuable, high-quality skins which are used to make shoes, purses and handbags. At one time over 300 000 skins were purchased annually from the natives. The females of this species bear about 30 live young directly into the water.

YELLOW-LIPPED SEA KRAIT
Laticauda colubrina

The Yellow-lipped Sea Krait is confined to tropical parts of the eastern Indian Ocean and the western Pacific Ocean. It attains a length of 1.5 m. It has venom fangs in the front of its upper jaw, and its poison is very effective, especially on fishes. However, as the snake is rather placid, man is rarely bitten. This sea snake spends its life in the sea, but the females come ashore to lay their eggs. Although they stay ashore for only a short time, many are captured by the natives who sell them for their palatable flesh. The young hatch out of soft-shelled eggs and immediately hurry to the safety of the sea. This species feeds mainly on small, slender fishes.

YELLOW-BELLIED SEA SNAKE
Pelamis platurus

The Yellow-bellied Sea Snake has a vast area of distribution, covering almost all the warm seas. It is very abundant in Indonesia, Malaysia and India. This snake usually reaches only one metre in length. It has a conspicuously compressed tail and normally has yellow and black coloration. It lives permanently in the sea, never coming ashore, and the female bears live young in the water. This beautiful, venomous sea snake is often thrown on to sandy shores by the waves, and as it cannot get back into the sea again, it often

Racerunner Eremias arguta

falls victim to birds of prey. The Yellow-bellied Sea
Snake is said to have caused the death of sailors swimming in the sea, and certainly its poison is strong
enough to have done so. On the other hand, this small
snake seldom bites. It feeds on small fishes.

GRACEFUL SMALL-HEADED SEA SNAKE
Microcephalophis gracilis

The Graceful Small-headed Sea Snake is a remarkable
reptile. When it is coiled, it looks like two intertwined
snakes. The front part of its body is extremely thin
and the head is tiny, while the hind part is very thick.
This sea snake inhabits the Indian Ocean and the
western Pacific Ocean. It never leaves the water and is
only found on land when it has been washed up by
the sea. Then it dies, for it cannot get back to the sea
again. The female bears live young in the sea. This
snake feeds on minute fishes of the eel family. Although it is extremely poisonous, it cannot bite man
because its jaws are small and do not open wide
enough.

GOLDEN-BANDED GOBI
Brachygobius xanthozona

The golden-banded gobi *Brachygobius xanthozona* inhabits Borneo, Sumatra and Java. It measures about

Yellow-lipped Sea Krait

Yellow-bellied Sea Snake

Graceful Small-headed Sea Snake

4.5 cm and has a cylindrical body. It is mainly found in estuaries, living near the bottom, and only occasionally swimming to the surface. The female usually lays 100—150 relatively large eggs under stones or among the tangle of aquatic vegetation. The male guards the eggs and provides them with a continuous supply of oxygenated water by whirring his fins. The fry swim near the surface, only later adapting to life on the bottom. This fish feeds on small invertebrates and plant material.

ARCHERFISH
Toxotes jaculator

The Archerfish is native to south-eastern Asia, the Philippines and the islands around Australia. This remarkable fish is about 25 cm long, is highly variable in colour, and lives in the brackish waters of mangrove swamps and river estuaries. The Archerfish is famous for its method of hunting its prey. It is able to spit a large jet of water from its mouth, hitting insects on leaves above the water or flying over the surface. An adult specimen can spit over a distance of about 4 m, causing an insect to fall into the water, where it

Golden-banded gobi Brachygobius xanthozona

Argus-fish

Archerfish

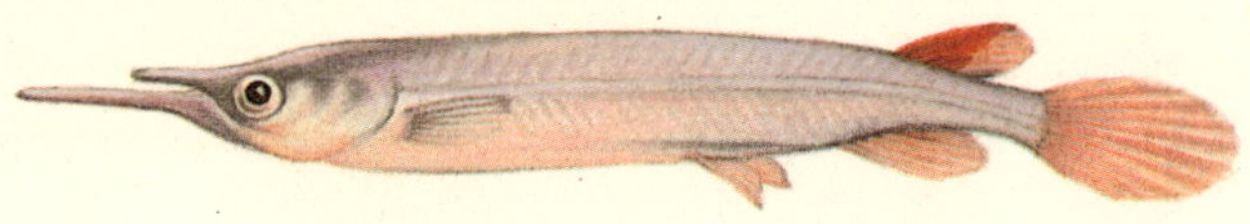

Halfbeak Dermogenys pusillus

is swiftly grabbed and swallowed. Archerfishes also feed on aquatic invertebrates. At first the young fishes swim in the water among the mangroves, catching crustaceans, molluscs and insect larvae. Only when they reach a length of 2—3 cm do they begin to learn to spit water skilfully.

ARGUS-FISH
Scatophagus argus

The Argus-fish reaches up to 30 cm in length and is highly variable in colour, with shades of green, blue and gold predominating. This fish is found in coastal waters and in the brackish waters of estuaries. It is distributed from India to Australia, and also occurs on Tahiti. It has fast-growing fry. This omnivorous fish feeds on algae, fine plant material, worms, molluscs, insect larvae and small crustaceans.

MUDSKIPPER
Periophthalmus koelreuteri

The mudskipper *Periophthalmus koelreuteri* is found along the coasts of the Indian and Atlantic Oceans, in tropical waters. It grows to a length of 15 cm and has mobile, protruding eyes. It is one of the few fishes that is able to leave the water. It uses its muscular pectoral fins to crawl over the mud of the mangrove swamps as the pools dry out at low tide. It can remain on land for some hours, the moisture-laden air preventing the skin and gills from becoming dry. Mudskippers can also leap distances of several metres, and they often climb on to a stone or a sloping branch, to observe the territories which they defend against their rivals. They are extremely wary, and when disturbed, dive into the water and hide in the mud. The Mudskipper

preys on insects, molluscs, worms and crustaceans, especially those stranded on the land by the tide, and often pursues its prey across the mud.

At spawning time, mudskippers build a funnel-shaped nest filled with water. The female lays her eggs in the lower part of the nest and guards them until they hatch. After a few days the fins of the young mudskippers become elongated. These fins are later used for crawling.

HALFBEAK
Dermogenys pusillus

The halfbeak *Dermogenys pusillus* is found in both fresh and brackish waters of Thailand, Malaysia and the Greater Sunda Islands. The female reaches a length of 8 cm, while the male is smaller. This strange fish looks as if its upper jaw has been cut off, for the lower jaw is considerably elongated. This adaptation enables the fish to catch prey moving above it. For this reason it stays near the surface, where it collects insects which have fallen into the water. It also catches gnat larvae. This species is viviparous, the female bearing 12—30 young which are about 1 cm long at birth. At first their jaws are of the same length, and only later does the lower jaw become elongated.

ROBBER CRAB
Birgus latro

The Robber Crab inhabits island shores and mainland coasts in the tropical regions of the Indian and

Mudskipper Periophthalmus koelreuteri

Robber Crab

Pacific Oceans. It measures up to 45 cm. Its coloration is variable, red, yellow or greyish specimens being seen. This land hermit crab has a soft, irregular abdomen covered by horny plates. It digs holes in the ground in which to rest, and it also hides in rocky crevices or in holes in trees. During the day, it forages for coconuts. It is able to climb palm trees, snipping off the nuts so that they fall down. Then the crab eats them on the ground. It is expert at cracking the nuts with its huge claws, using them like a pair of pincers. It removes the fibrous husk, and in the soft spot on the shell where the fruit germinates, the crab makes a hole through which it scoops out the pulpy contents,

using the tiny claws on its fourth pair of legs. The young hatch from eggs and live in the sea. As they mature they seek empty shells and shelter in them to protect their delicate abdomens. Older specimens forsake the shells and leave the sea. On land, in addition to coconuts, the robber crabs feed on other small crabs which they seize on the shore.

MOTH
Erasmia pulchella

The moth *Erasmia pulchella* is found in India and Assam. It is one of the largest and most beautiful species of its genus, having a wingspan of about 75 mm. During the day it flies slowly in open localities, frequently alighting on tall flowering plants.

ORANGE ALBATROSS BUTTERFLY
Appias nero zarinda

The Orange Albatross Butterfly is distributed throughout India and the adjacent islands. It is a very common butterfly, with a wingspan of 7 cm and an attractive yellowish or orange coloration. It has been observed to undertake long journeys similar to the migration of birds.

Moth Erasmia pulchella

Orange Albatross Butterfly

THE NORTHERN
AREAS

The vast coast of northern Asia is part of the territory of the Soviet Union. It is dotted with many rocky islands, bizarre rock forms and massive cliffs, carved out by the sea over millions of years. The steep rocks are normally inaccessible to men or predatory animals, and so they have become the refuge of scores of species of marine birds. At nesting time, the rocks swarm with birds, each pair occupying only the smallest nesting territory. When the nests are crowded close together, a great deal of fighting and stealing of nest material takes place. These sites are sometimes called bird mountains. Brünnich's Guillemot *(Uria lomvia)* is the most common bird, being found along the northern coast of Siberia, on Wrangell Island, on the coasts of the Chukchi and Bering Seas, and along the Kuril Islands and Sakhalin. Populations of these birds run into millions. The islands along the northern Asiatic coastline are also inhabited by the common Black Guillemot *(Cepphus grylle)*. The coasts of Kamchatka and Sakhalin provide nesting grounds for the Tufted Puffin *(Lunda cirrhata)*, a beautiful bird which digs nesting burrows in the clay soil of steep slopes near the sea. The coastal areas of northern Asia are also inhabited by many species of gulls.

In spring, northern species of ducks, geese and swans can be seen in large numbers in these regions. The King Eider *(Somateria spectabilis)* is found on the coast of arctic Asia, the male displaying beautiful coloration in the courtship period. Northern Siberia is the home of the Spectacled Eider *(Somateria fischeri)*, and Steller's Eider *(Polysticta stelleri)* also lives in coastal areas. The handsome Red-breasted Goose *(Branta ruficollis)* nests in the arctic tundra. It is currently protected, both in its nesting grounds and in its winter habitats in the region of the Caspian Sea. The total population of this species is estimated at 40 000. Another, more common species, the Brent Goose *(Branta bernicla)*, is also native to the arctic tundras of Asia. The Snow Goose *(Anser caerulescens)* is one of the most abundant species of wild goose, being found

mainly on the arctic coasts of North America. Around 1750, the Snow Goose was widely distributed in the Siberian tundras as well, but it was not observed after the beginning of the early 19th century. Snow geese used to overwinter in Japan, but they have not been seen there since the middle of this century. One reason for their disappearance is that when they moult they are no longer able to fly, and so cannot escape when they are hunted. Furthermore, snow geese nest on coasts which are easy of access, and Eskimos once hunted them, catching them by the thousand. A larger number of snow geese now inhabits Wrangell Island, and in recent years, as a result of increased protection, the birds have begun to return to their former nesting grounds in Siberia.

The Japanese or Manchurian Crane *(Grus japonensis)*, of the island of Hokkaido, has a similar history. In the second half of the last century, it occurred in large numbers on the island, and was regarded by the villagers as the herald of spring. In early spring, when the land was in some places still covered by snow, the cranes came to the northern regions and settled beside fast-flowing rivers which never froze over, in the vicinity of human settlements. The people had a high regard for the cranes and fed them on grain, and the cranes approached the settlements to feed, knowing that people would not hurt them.

At the beginning of this century, however, the numbers of cranes began to decline. This was brought about mainly by the reclamation of land formerly used as nesting grounds. It was soon realized that the Japanese Crane no longer lived in Japan, and this state of affairs continued for a long time. Recently, a flock of cranes again appeared in the vast swamps of Hokkaido, much to the joy of both zoologists and the local inhabitants. Once again the island was visited by the herald of spring. The Japanese Crane was immediately put under strict protection, and it is hoped that it will multiply and once more enliven the monotonous landscape.

In the arctic lands of Asia, rare carnivores can also be encountered. The largest is the Polar Bear, which is protected by law in many parts of its area of distribution, notably on Wrangell Island where about 250 of these animals live in a special reserve. Of the total world population of polar bears, 8 000 live in Asia.

The largest living deer — the Elk — abounds both in European northern regions and in the far north of Asia.

The vast northern regions consist of characteristic zones of tundra. Here the immense low-lying plains and rugged country are permanently frozen except during the brief arctic summer when the soil thaws through only to a depth of 30—50 cm. The water from the melted snow cannot soak into the soil, and so it forms thousands of shallow pools. These become the home of many plants and provide breeding grounds for a variety of insects. At this time, the young of the northern birds become fully fledged, and the birds migrate singly or in flocks from their cold native habitats to their winter grounds in the tropics of southern and south-eastern Asia. Many buntings, wagtails and plovers travel to India, Sri Lanka, Burma, Thailand or Vietnam. The Lesser or Pacific Golden Plover *(Pluvialis dominica)*, which nests in northern Siberia, wanders as far as the Sunda Islands and Australia, while the eastern Siberian populations of the Red-necked Phalarope *(Phalaropus lobatus)* migrate to the islands between southern Asia and Australia, and spend the winter there.

Tundras are covered by an endless carpet of mosses, especially those of the genus *Polytrichum* and *Dicranum,* by scant tufts of sedge, and by a few plants, such as species of the genus *Saxifraga* and *Chrysoplenium.* Thick layers of peat cover many areas, glittering in summer with a mosaic of pools and bogs. These waters attract many ducks, geese and swans. Willows of the genus *Salix,* often in the form of stunted, creeping shrubs, grow among the bogs, and white-

flowered shrubs of the genus *Dryas* grow close against the ground, their horizontal branches giving them the appearance of flat loaves. Arctic shrubs of the genus *Empetrum,* which have many branches and dense needle-shaped leaves, form thick pads, and in summer produce fragile pink blossoms which after a few weeks ripen into tiny fruits the size of blueberries. Junipers can be found even in the extreme north, though the shrubs reach only 30 cm in height. The Mountain Juniper *(Juniperus montana nana),* which has very short needles, abounds here. Blackberries are a feature of peat bogs, and the Cloudberry *(Rubus chamaemorus),* which is 10 cm tall and has white blossoms and reddish-orange berries, is common.

The Cranberry *(Vaccinium oxycoccus)* is a characteristic bog plant of the tundra, especially in arctic Siberia. This low, creeping plant has pale pink flowers. In late summer, its large, red berries measuring 10—15 mm across, provide valuable food for bears, foxes, small mammals and many birds. The berries can also be used to make preserves, wine and liqueurs.

To the south of the tundra an immense expanse of forest stretches right across northern Asia. This forest belt, called the taiga, gives rise to a rather monotonous landscape. At its northern limit birches, aspens, spruces, pines and larches grow.

The Siberian Larch *(Larix sibirica)* is a frequent tree of the area of taiga which ranges from the Kola Peninsula and the southern Urals across Siberia to the Sea of Okhotsk. This species has needles 3—4 cm long and cones measuring about 4 cm in length. Stunted forms of *Larix dahurica* with their distorted trunks are also common in the bogs, and in some areas there are stands of Scots Pine *(Pinus sylvestris)* and Arolla Pine *(Pinus cembra).* The latter species is the more important. In middle life, this pine has a dense, cylindrical crown, but later it spreads its branches, which are often irregular. The cones are ovoid or spherical, and up to 6 cm long. The large edible seeds are called pine nuts. When pressed, they yield high-quality oil.

Further south, more deciduous trees make their appearance in the taiga, and the plant cover gradually changes to mixed forests and then to deciduous forests with a thick undergrowth of bushes and tall herbs. Oaks, beeches and elms predominate in the deciduous and mixed forests.

Northern Asia is an area where water is abundant. Here flow massive rivers with enormous drainage basins. Largest is the basin of the Yenisey, which is 4 130 km long, and occupies an area of 2 707 000 square kilometres. The Ob, which is 4 070 km long, is another huge river, and so is the Lena, 4 320 km long, the waters of which feed Lake Baikal. All these rivers flow into the Kara Sea. The Amur, which measures 4 350 km, is the longest of the Siberian rivers flowing into the Sea of Okhotsk. One of the largest lakes in the world is also found in southern Siberia. This is Lake Baikal, which has an area of 30 500 square kilometres, and with a depth of 1 742 m is also the deepest lake in Asia. This lake plays an important part in the economy of the region. It provides drinking water for the large number of new towns which now surround it, and many species of fishes of great value to the fishing industry are found in its waters. Fish-processing factories are built along its shores. Lake Baikal is also the home of the fish-eating Baikal Ringed Seal, which is a protected species. Small streams and lakes are frequently inhabited by the Beaver, which is also common in the north of Europe.

In north-eastern Asia, particularly in Kamchatka, Sakhalin, and in northern Japan there are scores of active volcanoes. Smoke rises ominously from their tops, and steam hisses from openings in their sides. Earthquakes are also a frequent occurrence in this area.

The sea around Sakhalin is dotted with many small, rocky islands which become the temporary homes of sea lions. Huge herds of marine mammals gather every year and hundreds of them squeeze onto the flat rocky cliffs. As soon as the young become independent, the sea lions swim out to the open sea, but the young return to the islands again a few years later, as adults.

The islands are also inhabited by thousands of sea birds such as gulls and guillemots.

Spruces and pines are common in these areas and logging is carried out on a large scale. On Sakhalin, Giant Knotweed *(Polygonum sachalinense)* is widespread. It grows to a height of over 3 m, and has huge leaves, up to 30 cm long. Japanese Knotweed *(Polygonum cuspidatum)* grows on Hokkaido. It is also about 3 m tall and has hollow, branched stems.

In the subarctic zone, plants flower only in June and July, but they exist in great variety. The summers are pleasantly warm but humidity is high, and swamps and lakes are of wide occurrence.

EUROPEAN FLYING SQUIRREL
Pteromys volans

The European Flying Squirrel, related to the true squirrels, inhabits coniferous and mixed forests in Siberia, the coasts of China and Korea, and the Japanese island of Hokkaido. It reaches a length of 34 cm including the tail, which measures 12 cm. Flying squirrels have long, fine hair, and are variable in colour, ranging from silvery grey to yellowish-grey. They live in the canopy, leaping nimbly from tree to tree and covering distances of up to 50 m in gliding jumps, using their tails to control direction. In contrast, they are clumsy on the ground and rarely descend from the trees. During the day, they sleep in the trees, usually in holes vacated by woodpeckers, but sometimes in the nests of other kinds of birds. They are nocturnal, coming out to forage at dusk for nuts, seeds, berries, fruit, mushrooms and leaves. They also feed on insects, eggs and the nestlings of small birds.
The European Flying Squirrel lives in pairs or in family groups. After a gestation period of about 40 days, the female gives birth to 2—4 blind and naked young in a lined nest. The young open their eyes after 25—28 days. Although this species does not hibernate, it is not very active in winter. Its chief enemies are eagle owls and arboreal carnivores of the marten family.

SIBERIAN WEASEL
Mustela sibirica

The Siberian Weasel is widely distributed from Japan and Korea across southern Siberia to the Ural Mountains, but it also occurs in China and India. It is up to 48 cm long including the tail, which measures about 18 cm. The Siberian Weasel lives in thin woodland and in bush-covered and rocky localities. It hides in hollow trees, rocky crevices or under thick bushes by day, coming out to hunt at nightfall. It scurries silently on the ground, seizing small rodents, birds and reptiles, but in times of shortage, it makes do with worms and molluscs. It lives singly for most of the year, the male joining the female only in the courtship period, and then only for a few hours. Before giving birth to 1—2 young, the female lines the nest with moss, leaves and feathers. The gestation period lasts about 38 days. The young are born blind, opening their eyes after 28—30 days. They are suckled for 8 weeks and mature at the age of 10 months.

SABLE
Martes zibellina

The Sable is an inhabitant of the taiga in northern and eastern Siberia. It reaches a length of about 45 cm, of which 15 cm is made up of the tail. In its territory, this active carnivore covers over 15 km every night in quest of prey. It is also active by day, hunting for squirrels and small birds in the trees and for rodents on the ground, and it varies its extensive territories according to the availability of the small animals on which it feeds. It is solitary for most of the year, and pairs live together only briefly during the courtship period in July and August.
After a gestation period of 250—300 days, the female bears 3—4 young in her nest. Young sables are born blind, obtaining their sight after 30—35 days. They are suckled for 6—7 weeks, and they mature after 15 months.
The pelt of the Sable is very beautiful, the most valuable fur being blue-black in colour, with silver tips.

European Flying Squirrel

Every year trappers send hundreds of pelts for international auction in Moscow, though sables in Russia are now bred on farms.

PALLAS'S CAT
Felis manul

Pallas's Cat is native to an area extending from just east of the Caspian Sea to Tibet and western China. It is a beautiful wild cat, reaching a length of 70 cm, out of which 22 cm are taken up by the bushy tail. Its fur is about 6 cm long. It lives on the steppes, where each individual has its own territory. It can also be encountered among the mountains. During the day or at dusk, this cat follows regular paths in search of prey, marking its territory by scratching the bark of trees or bushes to reveal the white wood. This carnivore preys mainly on rodents, piping hares and fowl-like birds. In the courtship period, in February or March, the male seeks a mate and stays with her for a few days. After a gestation period of 65 days, the female bears 1—7 young in a sheltered place, either in a rocky cave or in a hole in a fallen tree. The kittens are born blind, opening their eyes after 2—10 days. They are suckled for 6—8 days and become independent at the age of 3 months. They are mature at 10 months.

USSURIAN or SIBERIAN TIGER
Panthera tigris altaica

The Ussurian or Siberian Tiger is the largest subspecies of tiger. The male stands over 1 m tall at the shoulder and reaches a total length of 375 cm, of which one third is made up of the tail, and he weighs over 280 kg. The Ussurian Tiger is indigenous to the Amur basin, north-eastern China and part of North Korea. At present, most of these massive carnivores live within Russian territory where they are protected. According to the latest reports, about 200 tigers live in this area. The Ussurian Tiger has extremely long thick hair which protects it against the cold, for the temperature here drops far below freezing point and

Siberian Weasel

Sable

Pallas's Cat

snow can fall to a depth of over a metre. Tigers live in vast forests, each individual having its own hunting ground where it covers scores of kilometres every day in search of prey. It hunts mainly during daylight and at dusk, but in places where it is disturbed it comes out at night. Despite its huge size, it is expert at stalking its prey. It crouches motionless waiting for the right moment to pounce, only the twitching tip of its tail betraying its excitement. The tail is also used by the female to give instructions to her cubs as they learn to hunt. This tiger kills small animals by gripping their necks, and larger ones by biting through their throats. It mainly hunts wild boars and deer, though it also catches wolves, or even sometimes a bear. This only happens when food is in short supply, for a tiger may very well come out second best in an encounter with a bear. In summer tigers catch small mice, frogs, birds, and even insects, and they eat forest fruits.

The Tiger drags its prey into the shelter of a thicket to devour it in peace. It can consume over 30 kg of meat at a sitting, and then go without food for several days. It covers the remains of the kill with twigs, leaves and moss. If it is unable to catch any prey, it can survive for two weeks or more without food.

In the breeding season tigers form pairs, rival males often fighting for a particular female. The pairs frequently stay together for a few weeks after mating has taken place. In early spring, after a gestation period of 105—108 days, the female gives birth to 2—4 young in a cave or underneath a fallen tree. The cubs are blind at birth, opening their eyes after 3—10 days. Their mother tends them carefully, licking them clean, massaging their bellies and playing with them. Young tigers leave the den for the first time when they are 40 days old, and begin to receive solid food at the age of 2 months. They continue to be suckled for 5—6 months, after which time their mother takes

them hunting. At 11—15 months, young tigers become independent and hunt larger prey on their own. They stay with their mother for 2—3 years. The males mature after 3.5 years, the females a year earlier.

The Ussurian Tiger likes to bathe and it is a good swimmer, easily crossing rivers in its path. It never climbs trees, but uses tree trunks to sharpen its massive claws. In the courtship period, the male roars loudly and also makes a hollow growling sound.

SIKA
Cervus nippon

The Sika is distributed in Manchuria, Korea, eastern and central China, the Ussuri region and Japan, occurring in several subspecies. In summer it is covered with large white spots which disappear when the deer grows its winter coat. This deer stands 120 cm tall at the shoulder and is 170 cm long. The male usually has antlers with eight tines. In the rutting season, which lasts from October to December, the bulls fight for possession of the herds of females. After a gestation period lasting 218—246 days, the female gives birth to a single young, or occasionally twins. The fawn hides in dense tussocks of tall grass or among bushes, while its mother stays on guard nearby and comes to feed it. It is suckled for about 7 months, although it is able to eat solid food from two months.

The Sika feeds on green plants, twigs, shoots and bark. In many regions this species was virtually exterminated in the search for the short immature horns of the young deer. These were often cut off complete with the cranial bone, to be of the greatest value. Sikas were killed in huge numbers as a commercial venture. The horns were in such great demand because a valuable medical substance, pantocrine, was obtained from them. Pantocrine was used mainly in Chi-

na, and it does actually have curative properties. It can be used to treat arteriosclerosis and rheumatism, and to stimulate the heart. In some areas, the Sika is kept in a half-wild state, and it is sometimes bred on farms. Such deer are not killed, their horns being expertly removed. In some countries, the Sika can be seen in deer parks.

ALTAI MARAL
Cervus elaphus sibiricus

The Altai Maral, which is a subspecies of the Red Deer (*Cervus elaphus*), inhabits the mountainous areas of the Altai and Tien Shan, and the area around Lake Baikal. It also lives in central Asia. The bull reaches a height of 160 cm at the shoulder and a length of 225 cm. The female is smaller and more slightly built. In spring, usually in March, the males grow new antlers which reach their normal size at the end of July or in early August. Then follows the rut, when robust adult males engage in fierce fights, and their loud, deep bellows can be heard far and wide. After each encounter, the victor takes the herd of females, and threatens any other rivals who venture too near. One male can be accompanied by 30—50 females, although he becomes solitary again when the rutting

Baikal Ringed Seal

season has ended. The female gives birth to a single young, or sometimes twins, in a shelter among thick grass and brushwood, after a gestation period of 225—262 days. On the second day after birth, the young deer becomes very active, but it remains in its shelter for a few days longer, while its mother comes to feed it. Suckling lasts up to 8 months, but the young deer is able to feed independently much earlier. The diet of this deer is made up of grass, leaves, bark, shoots, fruits, lichen and moss. Tigers, bears and wolves are the main enemies of this red deer. Like the Sika, this species is farmed for its immature horns which yield pantocrine.

BAIKAL RINGED SEAL
Pusa sibirica

The Baikal Ringed Seal lives in Lake Baikal, which is one of the largest lakes in Asia. It is a small species, reaching a length of only just over 1 m. It lives in large groups which often hunt together. These seals dive to a depth of 15—20 m, where they catch fishes, crustaceans and molluscs. When replete, they crawl onto the rocky shores and bask in the sun. In winter, when the lake freezes over, the males stay beneath the ice, and only pregnant females remain on the icy surface. The remaining seals come out onto the ice at the beginning of April and form groups of 100—200. When the ice melts, the seals swim together to the northern part of the lake, and settle on rocks projecting from the water. After a gestation period lasting 11 months, the female gives birth to a single young on the land, and stays with it continuously for the first few days, without taking any food herself. The cub, born with a thick coat of fur, is suckled for 5 weeks. Seal milk is nutritious and has a high fat content, so the cub grows quickly. When it is 3—5 weeks old it grows a new coat of stiff, coarse hairs which do not absorb water, and it is able to enter the water for the

first time and learn to hunt, beginning with crustaceans and molluscs, which are easily caught. After 2 weeks, the young seal is an excellent hunter. It matures after 3—5 years. The present population of the Baikal Ringed Seal is estimated at about 50 000.

BLACK-THROATED DIVER or ARCTIC LOON
Gavia arctica

The Black-throated Diver, or Arctic Loon, is found from the Urals to Lake Baikal. In the breeding season it settles on deep lakes, usually near the coast. The diver, which is about 65 cm long, builds its nest on islands in the lakes, preferably over the water, so that the birds can slip from the nest straight into the water. In April or May the female lays 2 spotted eggs in a scrape in the grass, usually unlined, and both partners undertake incubation for 28—32 days. The mother takes her brood to the water as soon as their down has dried, and they remain there until the young have grown their flying feathers. Divers leave the northernmost nesting grounds in mid-August. If the young are still unable to fly, they swim downstream and remain on the sea for a time before migrating south. In November they set out for their winter grounds, which in Asia are situated on the northern coast of the Caspian Sea. They return to their northern homeland between the beginning of March and the end of April. This species feeds on fishes, crustaceans, molluscs, worms and aquatic insects. The birds dive to a depth of 30 m, or even 45 m in search of prey.

SLAVONIAN or HORNED GREBE
Podiceps auritus

The Slavonian or Horned Grebe has an area of distribution which covers all of northern Asia up to latitude 70° N. It also lives in northern Europe and North America. It reaches a length of 33 cm and both sexes are alike in colour. In the southern areas of its distribution it also occurs on lakes in the mountains, being found in the Altai at a height of 2 300 m. The Asiatic populations spend the winter on the Japanese islands, in Korea, eastern China and along the Caspian Sea, staying in flocks close to the shore. They return to their nesting grounds on lakes and in swamps in May. A small lake is usually inhabited by just one pair, while a larger body of water may contain a sizeable colony. The nest is built in June, near the bank in a swampy area overgrown with vegetation. The female builds the nest from material brought by the male, mainly rotting pieces of water plant. The female lays 4—5, or occasionally up to 8, whitish elliptical eggs, and both partners incubate the clutch for 20—24 days. The adult birds carry their young on their backs on the water and nurse them carefully for several weeks. This species feeds on crustaceans, insects, molluscs, worms, tiny frogs and fishes, green plants and seeds.

BEWICK'S SWAN
Cygnus bewickii

Bewick's Swan is resident on small islands and on the coasts of northern Asia. In winter this swan, which is

about 122 cm long, is found around the Caspian Sea, the Aral Sea, and in western Europe. Eastern populations migrate to the southern Japanese islands and to eastern China, and in their winter grounds, the swans often gather in flocks of several hundred. In April, they return to their native habitats in swampy tundras, particularly to areas of shallow water and small

Black-throated Diver or Arctic Loon

Slavonian or Horned Grebe

Bewick's Swan

White-fronted Goose

young are looked after by both parents and can fly by the time they are 45 days old. This swan feeds on grass, shoots, seeds, berries, water insects, crustaceans and occasional small fishes and tadpoles.

WHITE-FRONTED GOOSE
Anser albifrons

The White-fronted Goose, which reaches a length of 76 cm, is native to coastal areas in the tundras of northern Asia. It also lives in North America and on the coasts of Greenland and Iceland. The populations of western Asia spend winter in large flocks in west and south-east Europe, while the eastern populations migrate to Japan and China. They return to their nesting grounds in May, and in mid-June return to the northernmost regions. A few pairs sometimes form a small colony. The nest is built by the female on a raised site, usually on a hillside terrace. The nesting cup is lined with grass and leaves, and when the female begins to sit, she envelopes the clutch in a thick layer of down. She incubates her 3—7 yellowish-white eggs for 26—30 days, while the male perches close by. The young are watched over by both their parents, and after fledging they join flocks which merge to visit damp meadows rich in lush vegetation, especially those on the banks of rivers and lakes. The White-fronted Goose feeds on grass, leaves, berries and seeds.

streams. The nest is situated in a raised spot near the water. It is built by the female alone, from moss, lichen, grass stalks and sticks. The cup is lined with a large quantity of down. In late May or June the female lays 2—3, or sometimes up to 5, yellowish-white eggs. The female incubates the clutch for 30—35 days while her partner stays nearby and keeps watch. The

Red-breasted Goose

RED-BREASTED GOOSE
Branta ruficollis

The attractively coloured Red-breasted Goose, which is about 55 cm long, has a relatively small area of distribution. Its nesting grounds are found in the tundra, stretching from the eastern coast of the Yamal Peninsula eastward across the delta of the river Dudinka to the western coast of the Taimyr Peninsula. The Red-breasted Goose migrates in flocks to regions south of the Caspian Sea and to the western coast of the Black Sea, and occasionally small groups reach central Europe. It returns to its nesting grounds after June 10, for until that time the tundra is covered by snow. It nests in small colonies usually of 5—7 pairs, though there may be more, the colonies often taking up residence near the nesting grounds of falcons or northern rough-legged buzzards. The nest is built by the female among small bushes and grass on a sloping river bank or on an island in a stream. On the Taimyr Peninsula the geese nest on rocks above rivers. The nest is lined with grass stalks, leaves and a thick layer of down. The female normally begins to sit on the complete clutch of 3—9 yellowish-white eggs around June 20, and the goslings hatch out after 25 days. The gander stays on guard nearby throughout this time. When the young have dried, their parents take them to the water. When danger threatens, both young and adult birds can dive to escape their enemies. The geese sleep on the water and also shelter in reed beds. The young are able to fly by August 20, and flocks leave early in September, by which time the adult geese have grown new feathers. The Red-breasted Goose spends only 93—113 days in its nesting ground. This goose feeds mainly on seeds of plants of the genera *Galium*, *Bolboschoenus*, *Potamogeton* and *Salicornia*, as well as on leaves and grass. During their migratory flights, they eat wild garlic in the steppes of Kazakhstan. These geese are preyed upon mainly by foxes, and the young are hunted by large species of gulls. The present population of the Red-breasted Goose is estimated at 50 000.

BRENT GOOSE
Branta bernicla

The Brent Goose occurs in several subspecies throughout the tundras of the arctic area. It measures about 61 cm in length. In northern Asia, this goose nests along the coasts and on the offshore islands. Populations from the eastern regions spend the winter mainly on the coasts of Japan and China, and populations from the west migrate to coastal areas of western Europe. Brent geese return to their nesting grounds in the first half of June, and immediately start to build their nests. The female scrapes out a shallow depression and lines it with plant material and later with down. Groups made up of several pairs often join colonies of large species of gulls or build their nests near the nesting grounds of raptors. The female usually lays 3—6 whitish eggs and sits on them for 24—26 days. The males soon leave to form independent flocks. The young are able to fly by early August. In summer these geese feed on grass, lichen, moss, crustaceans, molluscs, worms and aquatic insects, but in winter they exist mainly on seaweed.

Brent Goose

BAIKAL TEAL
Anas formosa

The Baikal Teal is an inhabitant of the Siberian area, ranging to Kamchatka in the east and to the Yenisei River in the west. This species, which is about 40 cm long, is a surface-feeding duck. In his nuptial plumage the male displays striking, gaudy colours. The female has a very dull, inconspicuous coloration, valuable since she incubates the eggs and rears the young. In autumn the Baikal Teal migrates to the Japanese islands and to south-east Asia, where it lives in company with other species of duck. It returns in late April or May. This teal is found on lakes, rivers and marshes with large areas of water. At the end of May, the female builds a nest in a depression on the ground, usually near the water, but sometimes as far as 100 m from it. She lines the depression with grass

Baikal Teal

Falcated Teal

and leaves, and envelopes the clutch of 6—9 pale greenish eggs in down. She incubates the eggs for 24—28 days. This species feeds on seeds, green plants and plankton.

FALCATED TEAL
Anas falcata

The Falcated Teal is distributed in eastern Asia, in the west ranging to Lake Baikal, and in the east to the island of Hokkaido. It measures about 50 cm. The male in his nuptial plumage is one of the most beautiful species of duck. Teals spend the winter in Korea, in the southern Japanese islands, and in south-eastern Asia. In these winter grounds, they live on inland lakes and rice paddies, or along the coasts, and they form pairs before returning in May. Then they inhabit lakes and rivers with banks densely overgrown with vegetation. The female builds the nest on the ground, near water. It is a shallow scrape lined with leaves or grass, and the clutch is enveloped in down. The female incubates 6—8 yellowish eggs for 24—26 days, and as soon as the newly-hatched young are dry, she takes them to the water. Young teals feed on small crustaceans, insects, molluscs and green plants. Adult birds also eat seeds.

MANDARIN DUCK
Aix galericulata

The Mandarin Duck reaches a length of 43 cm. The drake in his nuptial plumage is one of the most beautifully coloured birds in the world. This duck is native to the Ussuri region, Japan, the Kuril Islands and maritime regions of eastern China including Taiwan. It spends the winter in eastern China and Japan, where it is even found on lakes in Tokyo parks. It forms pairs before reaching the nesting grounds at the end of March, or in April. During courtship the male swims on the surface, raises his head, lays his bill on the water and displays his colourful ornamental feathers. The female builds a nest in a tree hollow or on the ground among dense vegetation near water. In May or June, she lays 7—12 whitish eggs and sits on them for 28—30 days. As soon as they have dried, the newly hatched ducklings jump out of the nest, often from a height of well over 10 m, and fall either into the water or into tall soft grass which acts as a cushion. They are cared for by their mother. The Mandarin Duck feeds on insects and insect larvae, crustaceans, molluscs, green plants and seeds.

KING EIDER
Somateria spectabilis

The King Eider, which attains a length of 56 cm, is a denizen of the arctic coast of northern Asia. The female is inconspicuously coloured, being predominantly brownish, while the male in his nuptial plumage boasts a rich display of colours. These birds are never found more than 50—100 km inland from the coasts. In some areas the King Eider is very common, and the total Asiatic population is estimated at 1—1.5 mil-

Mandarin Duck

lion. The populations of Asia, North America and Greenland together comprise 4—5 million birds. In winter flocks of eiders 100 000 strong frequent the south-east of the Barents Sea, the western coast of Novaya Zemlya and the Bering Sea in the east. Small groups of eiders occasionally reach northern Europe. The birds begin to return to their nesting grounds at the end of March, but most of the flocks arrive in May, and the populations from the New Siberian Islands fly in during June. In the breeding season eiders live in pairs. The female makes a depression about 4 cm deep with her body and lines it with pieces of plant material, and with feathers, which protect the clutch from the cold. She lays 3—7 greenish-grey eggs and incubates them for 25—28 days, not leaving the nest even on the approach of man. When the newly hatched young have dried she takes them to the sea, often caring for other chicks who have lost their mothers as well. At the end of summer, the families move to quieter waters in lagoons, bays and river deltas. Young eiders are able to fly when they are 40—45 days old. The families then gather in flocks and leave the nesting grounds.

The whole nesting season from arrival to departure lasts only 80—85 days. This species feeds on mussels, midge larvae, sea urchins, echinoids, small fishes and crustaceans, particularly spider crabs. Eiders also feed on seeds, preferring those of the plant *Ranunculus pallasi.* They are preyed upon by polar foxes, snowy owls, falcons and large gulls, which attack the chicks.

SMEW
Mergus albellus

The Smew is found in the taiga throughout northern Asia. This duck is about 40 cm long and nests on lakes and rivers in coniferous forests. Several scores of pairs may settle on a large lake, arriving at their nesting grounds in April or in early May. The female seeks a suitable hollow in which to build her nest, while the male either follows her or waits nearby in the water. The female lays 6—11 cream-coloured eggs in a hole in a tree, usually about 2 m above the ground. She envelopes the clutch in down and sits on it for 30 days. The young jump out of the nest 24—36 hours after hatching and follow their mother to the water. When the brood has fledged, families gather to form flocks and leave for western and southern Europe, India and eastern Asia. Here they live on large

King Eider

rivers and lakes, or occasionally along the coasts. The Smew feeds on insects and insect larvae, molluscs, crustaceans, worms and small fishes.

STELLER'S SEA EAGLE
Haliaeetus pelagicus

Steller's Sea Eagle is one of the largest of the raptors. It reaches a length of 90 cm, has a wingspan of 250 cm and a weight of 6—9 kg. Sea eagles mature after 4—5 years when they moult into their attractive adult coloration. The breeding area of this eagle covers the coasts of the Bering Sea and the Sea of Okhotsk, Kamchatka, Sakhalin and the adjacent islands. In winter it is found along the coasts of Korea, Kamchatka and the Japanese islands. On the shores of Kamchatka it is partially resident. Adult sea eagles have white shoulders, a white tail, and a large, conspicuous yellow beak. The two sexes are alike in colour. Steller's Sea Eagle inhabits rocky seashores or river valleys near the sea, where there are tall trees. Pairs arrive at their nesting grounds early in March. At this time their nuptial flights can be seen, the partners soaring up,

Smew

gliding, plummeting down and taking to the air again. The clear, loud cries by which the birds greet one another can also be heard. Sea eagles usually build their nests on ledges among steep rocks or in trees, especially poplars or tall birches. The nest is a huge construction, up to 1.5 m high and about 2.5 m across. It is made of large branches and is lined with twigs and feathers. At the end of April the female lays 1—3 white eggs with a greenish sheen. Both partners incubate the eggs but the female spends more time on the nest. The chicks hatch out after 40 days or so, depending on the weather. They stay in the nest for 10 weeks and then take their first flights. Both parents feed the chicks bringing them fishes and various water birds. These birds also prey on small mammals and often collect dead animals cast up on to the beach by the sea. Steller's Sea Eagle returns to the same territory for many years.

NORTHERN ROUGH-LEGGED BUZZARD
Buteo lagopus

The Northern Rough-legged Buzzard is distributed throughout the arctic areas of Asia, America and Europe. It reaches a length of about 60 cm and has a wingspan of about 140 cm. It has characteristically feathered legs and a conspicuous dark spot on each wing, which shows when the bird is in flight. Individual buzzards are, however, highly variable in their coloration. The adult birds form pairs for life, but in au-

tumn they merge in large flocks to migrate to temperate zones where they spend the winter. In spring they return to their nesting grounds and on arrival they form pairs again and occupy their territories. The nest is made of branches, dry grass, blueberry twigs and lichens, and it is situated on the ground or on a rocky outcrop, or in a low tree. In May or early June, the female lays 2—7 eggs, the number depending on the availability of small rodents within the territory. The eggs are usually greenish or bluish with ochre, grey or violet patches, covered with russet spots. She incubates the clutch for 31 days, while the male stays nearby on guard. The young do not all hatch at the same time, for the female begins sitting when the first egg is laid. The male brings food to his mate, who passes it on to the young, and after 40—45 days the young buzzards are able to leave the nest. This species feeds mainly on lemmings but it also catches birds. In winter, voles constitute about 90 per cent of their food. Buzzards cruise low over the ground in search of their prey, often hovering in one spot, and they sometimes perch on stones near the entrances to the burrows, waiting for the rodents to emerge.

WILLOW GROUSE or WILLOW PTARMIGAN
Lagopus lagopus

The Willow Grouse, or Willow Ptarmigan, is widespread in several subspecies throughout the tundra of the northern hemisphere, and in Scotland and Ire-

land. It is very common in the tundras of northern Asia. It measures about 40 cm in length, and the sexes differ in colour. In the nesting season, the male is chestnut brown with black streaks, white wings and a black tail, while the female is predominantly fawn in colour with dark striping. The male of the subspecies found in Scotland is chestnut brown all over including the wings, and its plumage becomes a lighter shade in winter. Grouse living in the tundra are white in winter except for a black tail, which is covered above with white feathers. Both the legs and toes of this fowl are feathered, a feature which enables the birds to walk in the snow without sinking. In winter the birds of the Asiatic tundras often cover distances of up to 250 km as they fly to more southerly, forested areas. They return in spring to their nesting grounds, usually travelling in flocks. As soon as the snow melts the grouse scatter in the countryside and the courtship period begins. The males take to the air, fly about 2.5 m above the ground, soar sharply to 15—20 m and then plummet down. They accompany their flight with a bubbling call. In May or June the female builds a nest under a bush or among stones. There are sometimes more than 30 nests in one square kilometre, for this species is very abundant. The nest consists of a scrape lined with stalks and leaves. Here the female sits on her clutch of 5—12 yellowish, brown-spotted eggs for 24 days. When they are dry, the young leave the nest and their mother takes care of them. The Willow Grouse feeds mainly on plant material. In March and April catkins form 50 per cent of its diet, but the birds also peck green leaves, shoots and blossoms and they collect seeds. In summer they feed largely on berries. Young birds also eat insects and insect larvae.

SIBERIAN SPRUCE GROUSE
Falcipennis falcipennis

The Siberian Spruce Grouse is found in the Trans-Baikal region, along the Ussuri, and on Sakhalin. It attains a length of 50 cm. The female is predominantly chestnut brown with dense grey spots. This grouse lives in deciduous, mixed or sometimes coniferous forests in the mountains at an altitude of about 1 500 m. It settles mainly in localities rich in wood strawberries and cranberries, preferring birch and pine woods. It forms pairs in spring, but the male does not help to build the nest or rear the young. The female lays 6—8 eggs in a nest situated on the ground, at some time between May and July. The eggs are pale brown with

Willow Grouse or Willow Ptarmigan

Siberian Spruce Grouse

dark brown spots. The nest is usually situated near a tree trunk and is lined with twigs, grass stalks and leaves. The female sits on the clutch for 25 days. The Siberian Spruce Grouse feeds on berries, seeds, green plants, insects and insect larvae, and on other small invertebrates.

SIBERIAN WHITE CRANE
Grus leucogeranus

There are only two known nesting grounds of the Siberian White Crane in Siberia. One is in the north-west in the drainage basins of the rivers Konda and Sosva, and the other is in the north-east along the river Indigirka. The Siberian White Crane is a magnificent, long-legged bird reaching a length of about 140 cm. It inhabits large inaccessible swamps containing lakes with low banks overgrown with vegetation, and bogs in thin woodland. In mid-April its pleasant trumpet-

Siberian White Crane

ing call, more delicate and more melodious than that of other species of crane, can be heard in remote localities. The nest is built on a dry site and is made of sticks, grass stalks, reeds and moss. Both partners incubate 2 yellowish-green, dark spotted eggs for about 30 days. The young are active as soon as they are dry, and run around accompanied by their parents. After fledging, which takes place early in August, the young remain with the adults. The families then form small groups and lead a nomadic life, wandering far from their nesting grounds. In autumn they travel to their winter grounds. The western populations migrate to the southern coasts of the Caspian and Aral Seas, while the eastern population fly south to south-east and eastern China. At one time up to 300 cranes could be seen in a single wedge-shaped formation, but now this species is rare. The Siberian White Crane feeds on small fishes, frogs, reptiles, small rodents, insects, green plants and seeds.

JAPANESE or MANCHURIAN CRANE
Grus japonensis

The Japanese or Manchurian Crane, which is about 140 cm long, lives in the coastal area of the Ussuri region, and in adjacent parts of Manchuria, China and North Korea. It also occurs in the south-eastern tip of the Japanese island of Hokkaido. Cranes always seek swampy, inaccessible localities. They spend the winter in North Korea and along the lower reaches of the massive Yellow River in China. They arrive at their nesting grounds in the second half of March and leave between the end of August and November. On arrival, the cranes stay at first near rivers, choosing the same site every year, and usually living in small flocks. They form pairs early in April and then fly into the vast swamps, seeking places where the snow has already melted. Each pair chooses an area in which to nest, and defends it until it is time to leave. The nesting territory is extensive, the nests being several kilometres apart. During the nesting season the cranes make loud trumpeting sounds, which can be heard as far as 4 kilometres away. These birds are extremely cautious in the breeding season, and in open country they never allow anyone to get nearer to them than 200 m. The nest is built in the swamps and is made of reed stems, grasses and twigs. It measures about 1 m across. The female usually lays 2 spotted eggs. Both parents share in incubation for a month, and later

Japanese or Manchurian Crane

watch over their young. Japanese cranes mature after 5 years, when they begin to breed. Their diet is similar to that of the Siberian White Crane.

GREY or BLACK-BELLIED PLOVER
Pluvialis squatarola

The Grey or Black-bellied Plover is native to the tundras of the coastal areas of northern Asia and North America. The female resembles the male, but is brown below, with tiny white spots. For most of the year both partners have inconspicuous plumage, predominantly brown and grey spotted, with a whitish abdomen. In autumn, flocks of these birds migrate to Africa, south-east Asia, Australia and New Zealand. They cover incredible distances every year, although they are only about 28 cm long. They return to their nesting grounds in pairs and find dry sites in which the female can make her nest. This consists of a depression in the moss, lined with dry grass stalks and lichens. The complete clutch of 4 brownish eggs with

Grey or Black-bellied Plover

Lesser or Pacific Golden Plover

LESSER or PACIFIC GOLDEN PLOVER
Pluvialis dominica

The Lesser or Pacific Golden Plover is a bird of the tundras of America and of Asia, south to the Ussuri region. The female is similar to the male in appearance, but she usually lacks the pale stripe along the sides of the neck. Outside the nuptial season, this bird, which is about 25 cm long, has a greyish mantle, spotted with black and brown, and a white breast and abdomen. In autumn the Asiatic populations migrate to south-east Asia, the eastern coast of India, the Sunda Islands and the Australian coast. A few individuals may even wander as far as western Europe. They return to their nesting grounds again in May or June. Here the birds frequent swampy tundras and the edges of the taiga. The female makes her nest in a depression on the ground, lining it with small twigs and stalks. Both parents sit on the clutch of 4 pale brown eggs with russet and black spots for 27—28 days. When the young are dry, which is within a few hours of hatching, the adult birds take them out. This plover feeds on water insects, molluscs and crustaceans, and on seeds, green plants and berries. When searching for food, it scurries swiftly on the ground, often stopping abruptly.

RUDDY TURNSTONE
Arenaria interpres

The Ruddy Turnstone has an area of distribution covering the coast of northern Asia, Europe and North America. It is about 23 cm long. The female has a brown-spotted breast and an incomplete black stripe across her chest. Outside the breeding season, the birds lack the striking black and brown wing pattern. In autumn the birds of northern Asia migrate to India and Sri Lanka, while the populations from eastern Siberia cross Japan to the Pacific tropics, and the birds of western Asia fly across Scandinavia to Equatorial Africa. The Ruddy Turnstone travels at an incredible speed, one specimen marked by ornithologists covering 820 km in 25 hours. The birds leave their native land between the end of July and September, returning to their nesting grounds in April and May, or in June if they live in the northernmost regions. They settle in the same site every year. Courting males chase one another and engage in fights. The nest is built close to water on rocky islands and is

dark brown or blackish spots is laid at some time between mid-June and early July. Both partners undertake incubation for 23 days. The young leave the nest 1—2 days after hatching to roam in the neighbourhood. When danger threatens, they press themselves to the ground among the grass and stones. The parents provide shelter for them under their wings for about 14 days. When the brood is fledged, the families gather to form flocks. This plover feeds on insects, mainly beetles and gnat larvae, and also on molluscs, seeds and berries.

lined with dry plant material. In June the female lays 4 greyish-green to pale brown eggs with grey and brown spots. Both partners take part in incubating the eggs for 22—24 days, but towards the end the male takes over, and he sometimes tends the young without help from the female. The Ruddy Turnstone seeks its prey under stones or shells in shallow water. It turns them over with its bill and catches insects, molluscs, worms and crustaceans sheltering underneath. It also eats seeds and seaweed.

POMATORHINE SKUA or POMARINE JAEGER
Stercorarius pomarinus

The Pomatorhine Skua, or Pomarine Jaeger, is indigenous to the tundras of Asia and North America. This species reaches a length of about 52 cm and also occurs in a black form. This skua inhabits maritime regions and areas of tundra containing lakes and slow-running rivers. In the nesting season it lives in pairs or in small colonies. The birds build a nest on the ground, in the moss or among tussocks of grass, usually in June and July. The female lays 2—3 eggs, greenish or brownish-olive with dark spots. Both partners sit on the clutch for 24—28 days and together they watch over the chicks. The young birds stay in the nest for about 7 weeks and are fed by their parents. They often leave the nest while they are still unable to fly, hiding in the grass in the neighbourhood. The Pomatorhine Skua hunts for fishes, molluscs and crustaceans, but it also takes the eggs of other birds. It often pursues gulls until they disgorge their prey, skilfully seizing it in flight. They also feed on the remains of large animals left on the beach by the tide.

BLACK-TAILED GULL
Larus crassirostris

The Black-tailed Gull is distributed along the coasts of the Japanese islands, the Kuril Islands, southern Sakhalin, north-eastern China and south-eastern Russia. In winter it can be seen in regions ranging from southern Kamchatka to Taiwan, and sometimes as far south as Hong Kong. This gull follows the boats of fishermen, waiting for the remains of fish to be thrown into the sea. It also hunts various marine animals, cruising above the surface in search of prey, and it takes the eggs and nestlings of other birds. In the nesting season, which begins in mid-May, the gulls

Ruddy Turnstone

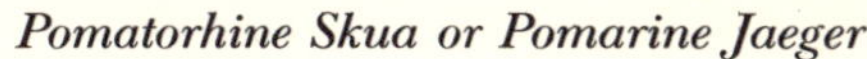

Pomatorhine Skua or Pomarine Jaeger

live in large colonies, building nests of dry grass on the ground or on rock ledges. The female lays 2—3 eggs, densely covered with brown, grey and violet spots, and both partners share incubation for 28 days. A few days after hatching, the young scatter and hide in the neighbourhood. The adults take food to them, and the young gulls begin to fly when they are 40 days old.

URAL OWL
Strix uralensis

The Ural Owl is found throughout the temperate zone ranging from northern Europe across Asia east to Korea, Sakhalin and the Japanese islands. It is about 60 cm long, and has a wingspan of 115 cm. In its native habitats the Ural Owl is resident, migrating to more southerly regions only in harsh winters. It lives mainly in coniferous forests, though it also settles in parks and overgrown gardens. It usually hunts after nightfall, but it also comes out on overcast days. It preys chiefly on squirrels, or on birds up to the size of a grouse, pigeons forming the bulk of its diet. Early in spring, often starting in March, the owl builds a nest in a hole in a tree or in a nest vacated by raptors, usually 10—20 m above the ground. The female lays 2—6 white eggs and sits on them usually for 27—29 days, though in very severe weather, she may need to sit for 34 days. The male brings food for her and the young for the first few days after hatching. He passes the prey to his mate and she divides it into portions for the chicks. The owlets begin to fly when they are about 50 days old.

GREAT GREY OWL
Strix nebulosa

The Great Grey Owl, which measures about 68 cm, ranges from northern Europe across Siberia to northern Sakhalin, and it also inhabits the western coast of Canada. In some localities, such as Yakutsk, it is one of the most common owls. It remains in its habitats even in winter, mainly frequenting vast coniferous forests, though in eastern Siberia it even inhabits deciduous woods. It hunts squirrels, lemmings and birds at night, though in the nesting season it hunts during the day as well. A courting male makes short hooting sounds, while the nesting female has a call

Black-tailed Gull

Ural Owl

not unlike that of cuckoo. This owl usually finds a nest
vacated by birds of prey, about 6—10 m above the
ground, and lines it with fir tree twigs. A pair uses the
same nest for many seasons. In April the female lays
1—7 eggs, the number depending on the availability
of small rodents for food. She incubates the clutch for
4 weeks, while the male feeds her. The young are
reared in the same manner as the young of the Ural
Owl.

ORIENTAL CUCKOO
Cuculus saturatus

The Oriental Cuckoo, which is about 33 cm long, has
a wide area of distribution ranging from the southern
Urals to eastern and south-eastern Asia and the Himalayas. In autumn it migrates to India, Malaysia, the
Sunda Islands, the Philippines and Australia, returning to its native land in spring. This cuckoo lives in
coniferous forests and in birch woods, and it can also
be seen in the mountains and on scrubland steppes.
In the breeding season this cuckoo produces monotonous sounds like the call of a hoopoe. The Oriental
Cuckoo is a brood-parasite, the female laying her eggs
in the nests of other birds, such as flycatchers, shrikes
and white-eyes. The eggs are white or pale brown
with dark brown and fawn spots. They are incubated
by the foster parents who feed the young cuckoo on
insects and insect larvae. The female lays about 15
eggs in one season, each in a different nest.

ASIAN BROWN FLYCATCHER
Muscicapa latirostris

The Asian Brown Flycatcher is native to north-eastern
Asia including Japan and Sakhalin. It also occurs in
the Himalayas. This bird, which is about 14 cm long,
inhabits deciduous forests in river valleys, seldom
settling in coniferous forests. In Japan, it can often be
seen in parks. It spends the winter in India, Burma,
Thailand and the Greater Sunda Islands, returning to
its nesting grounds from late May to mid-June. Soon
after arrival, it builds its nest on a horizontal branch
of a deciduous tree, about 1 m from the trunk. The
construction is bowl-shaped, and measures 10 cm a-
cross. It is woven from grass roots and a great quantity
of moss. The walls of the nest are usually covered by
lichens and small pieces of bark from the tree in
which the nest is situated. In this way the nest is per-

Great Grey Owl

Oriental Cuckoo

Asian Brown Flycatcher

Siberian Rubythroat

fectly masked. The nesting cup, which is about 3.5 cm deep, is lined with pieces of leaf. The clutch of 4—6 white eggs with greenish-grey spots on the broad end is incubated usually by the female for 12—14 days. The young are fed on insects and insect larvae caught by their parents. They leave the nest after 2 weeks, but the adult birds continue feeding them for another 10 days.

SIBERIAN RUBYTHROAT
Luscinia calliope

The Siberian Rubythroat has an area of distribution which covers the whole of Siberia as far as Kamchatka, and it also occurs on Hokkaido and in an isolated locality in south-western China. This bird reaches a length of 14 cm. The female is brownish in colour, with a white throat. In autumn the Siberian Rubythroat sets out on the long journey to its winter grounds, which are situated in south and south-east Asia and the Philippines. It usually migrates at night. It returns to its nesting grounds in spring, usually at the end of May. It settles chiefly in the taiga, preferring the fringes of forests, with an open aspect over meadows or rivers. On arrival, the male immediately begins to announce his presence by means of a melodious, repeated, whistling call. The female reaches the nesting grounds a few days later. The nest is built in a thicket on the ground, or sometimes among the branches of bushes about 50 cm above the ground. It is made of fine stalks, plant fibres and mammal hair, and it is not lined. It is usually spherical in shape, and has a side entrance. In June or July the female lays 3—6 eggs, greenish-grey in colour with russet spots and streaks. She incubates them for 14 days. The young are fed on insects, insect larvae, spiders and worms by both their parents.

RED-THROATED THRUSH
Turdus ruficollis

The Red-throated Thrush inhabits central Siberia and is found up to a height of 2 200 m in the Altai Mountains of Mongolia. The subspecies *Turdus ruficollis atrogularis* of the west of the area of distribution has a black throat. The female is brown with a white throat, and her breast is covered with black and white spots. This common songbird reaches a length of 23—26 cm. In autumn it migrates to the Himalayas, western China, Burma, northern India and Turkey. In the south the thrushes are not migratory but roam the countryside. They inhabit all types of forest but always live near rivers, as they like to bathe in shallow water close to the banks. The nest is built low down in a tree or occasionally on the ground. It is made from grass stems, and is lined with clay and fine grasses. The clutch of 4—7 greenish eggs densely covered with russet spots is incubated by the female for 2 weeks. The young are fed by both parents. The Red-throated Thrush feeds on small molluscs and insects, including aquatic species which the birds catch in shallow water. These thrushes also prey on spiders and earthworms, pecking them from the grass in the late afternoon.

DUSKY THRUSH
Turdus naumanni

The Dusky Thrush ranges throughout Siberia except for the most westerly parts, and reaches Kamchatka and Sakhalin in the east. It is about 25 cm long. The female is brownish in colour, with a black breast. This thrush spends the winter season in an area stretching from north-eastern Mongolia and northern China to south-eastern Asia and north-eastern India. It occasionally wanders as far as western Europe. During migration it travels in flocks of 10—30. In May it returns to its nesting grounds in forests and bush-covered localities. In late May or early June, it builds a nest of grass stalks, and plasters the inside of the walls with clay. The construction is situated in a bush or in a low tree. The female sits on the clutch of 5 greenish eggs with dark brown and russet spots for 12—14 days. Both parents care for their young and feed them for 2 weeks on the nest and for a further 10 days after fledging. The Dusky Thrush feeds on insects, spiders, worms and small molluscs. During the migratory flights and in their winter grounds, the thrushes also eat berries.

SIBERIAN THRUSH
Turdus sibericus

The Siberian Thrush is native to central and eastern Siberia, ranging in the east to Sakhalin and the northern Japanese islands. It is about 23 cm long. The female is brown with dark wavy lines below. In autumn these thrushes migrate to south and south-east Asia, sometimes flying as far as Borneo and Java. They return to their nesting grounds in May, settling along the edges of brushwood forests and in the taiga. They prefer to establish themselves close to rivers or lakes, where they can forage on the banks. They prey on insects, worms, spiders and molluscs, and in autumn they find berries. The nest is built in a bush or low tree, usually about 2—2.5 m above the ground. The

Red-throated Thrush

construction is similar to that of other thrushes, and is lined with fine grass and leaves. Building is carried out by the female, while the male brings the nest material. The female lays 4—5 eggs, usually beige in colour, with brown and grey spots, and she incubates them for 14 days. The young are fed by both parents.

Dusky Thrush

Siberian Thrush

Azure Tit

Scarlet Grosbeak

AZURE TIT
Parus cyanus

The Azure Tit is an inhabitant of the temperate zone of Asia, ranging in the south to southern China, and it also occurs in eastern Europe. This charming bird, which is only about 13 cm long, frequents dense forests containing thick undergrowth, and overgrown banks of rivers and lakes. In autumn it gathers in flocks of up to 100 birds, and wanders far from its nesting grounds. It ceases to be timid at this time, and appears in parks and gardens in towns and villages. It returns to its nesting grounds by mid-April, and pairs impatiently fly around in search of suitable nesting holes, while the males sing loudly. The nest is made of moss, dry grass and hair and is often built in a hollow vacated by woodpeckers, 2—4 m above the ground, or in a hole in a rotten stump. The clutch of 9—11 whitish, red-spotted eggs is incubated by the female for 13—14 days. The large brood is fed on tiny insects and larvae, mainly caterpillars, and spiders caught by the parents. The young leave the nest after 17—20 days and take their first flights. The adult birds continue to feed them for another 10 days.

SCARLET GROSBEAK
Carpodacus erythrinus

The Scarlet Grosbeak is widely distributed throughout northern Asia, except in maritime regions, and it also lives in north-eastern Europe. It can also be found in mountainous areas up to a height of 3 700 m, from the Caucasus to the Himalayas and eastern China. It reaches a length of 14.5 cm. The female is greenish-brown above and has a brown-spotted throat and breast, and a greyish-white abdomen. This grosbeak lives on the edges of forests, in the taiga, and in vast bush-covered areas, as well as in parks and gardens. It prefers sites near rivers. In autumn it migrates to northern India and south-eastern Asia, but it is resident in its more southerly areas of distribution, descending to low-lying land from its mountain habitats. It returns to its nesting grounds in May. The nest is built among dense branches of bushes or trees, usually 2—2.5 m above the ground. The construction is made of grass stalks, roots and tiny twigs, and the nesting cup is lined with horsehair and fine plant material. In early June the female lays 3—6 eggs, which are bluish-green, covered with dark brown and blackish-violet spots, particularly around the broad end. She

incubates the clutch for 13—14 days. The young are
fed by both parents and leave the nest when they are
14—17 days old. The Scarlet Grosbeak feeds mainly
on seeds, berries and insects, which the adults bring
to the newly hatched young.

WHITE-WINGED CROSSBILL
Loxia leucoptera

The White-winged Crossbill is found mainly in central and eastern Siberia, but it also occurs in north-
eastern Europe and in Canada and Alaska. It reaches
a length of 15—18 cm, and the female differs considerably in colour from the male. She is greenish-olive,
with a yellow rump and two white stripes on her
wings. In western parts of northern Asia, this bird
lives in the coniferous taiga, while in the Trans-Baikal
region and in the areas around the Amur, it is found
in deciduous forests. It feeds mainly on seeds from
cones, prising them out with the unusual crossed tips
of its beak, and on seeds of deciduous trees. In summer, its diet is supplemented by insect larvae and beetles. Before nesting, flocks of crossbills travel to localities where a variety of ripening seeds is available. The
nest is built by the female in trees, chiefly conifers,
about 3—5 m above the ground. It is made of dry
twigs, and the nesting cup is lined with fine bark and
hair. The male accompanies the female as she builds.
In July she lays 3—4 eggs, bluish with blackish-
brown dots and spots. She incubates them for 14—16
days, and the male feeds her during this time. The
newly hatched young have straight bills, which become crossed only after about 3 weeks. The chicks
leave the nest after 2 weeks, but the adult birds feed
them for another 10 days. After the brood has fledged,
crossbills roam the countryside in flocks.

PINE BUNTING
Emberiza leucocephala

The Pine Bunting inhabits the whole of Siberia as far
as the Amur and Sakhalin, and it also lives in an isolated area in northern China. It is resident in the
southern regions, but it migrates from the other localities at the end of August to spend the winter in central and eastern Asia, where it settles in open country.
It returns to central Siberia in mid-April, and to northern regions and to the Altai Mountains, where it oc-

White-winged Crossbill

Pine Bunting

curs at heights from 1 300—2 000 m, in June. Its habitats are thin, mainly coniferous woods. The nest is
built on the ground, in a tuft of dry grass or under
a bush. It is a shallow scrape lined with dry stalks or
horsehair. The nest is built by the female, who sits on
the clutch of 4—5 greyish-white eggs patterned with
brown spots and thin lines for 12—14 days. She is occasionally relieved by the male for a short time. At the
age of 2 weeks young buntings leave the nest, but they
continue to be fed for a few more days. The Pine Bunting feeds on the seeds of various plants, and in the
nesting season it also eats insects, insect larvae and

Siberian Jay

spiders. The length of this bird averages 16.5 cm. The female lacks the white nape and the white pattern on the cheeks and throat.

SIBERIAN JAY
Perisoreus infaustus

The Siberian Jay is distributed in northern and temperate Asia, and in northern Europe. It is not found in coastal areas or in Kamchatka. It reaches a length of 30 cm and both sexes are similar in colour. This jay remains in its native habitat even in the harshest winters, living mainly in coniferous forests in the taiga. After the broods have fledged, it sometimes moves to birch woods, and in winter it visits human settlements. Pairs nest in coniferous forests, where they build nests among branches 2—6 m above the ground. The foundation is made from dry twigs. These are covered by grass stalks and lichens, and the nesting cup is lined with hair and feathers. Sometimes in March, but usually in April, the female lays 3—5 greenish eggs covered with closely spaced grey and greyish-violet spots. Incubation is shared by both partners for 17—19 days. In the first week after hatching, the female stays with the chicks and keeps them warm, while the male brings food for the whole family. The young leave the nest when they are 23 days old. The Siberian Jay preys on insects, small birds and their eggs, and reptiles. It also feeds on the seeds of coniferous trees and on berries.

286

LONG-TAILED LIZARD
Takydromus amurensis

The long-tailed lizard *Takydromus amurensis* inhabits coastal areas of eastern Asia. It grows to a length of 15 cm, and has a very long tail, usually longer than the body. It is found mainly among cedars, though it can also be encountered in oak forests, living in holes or cracks in the bark or under fallen trees, and feeding on insects and larvae. The female lays 2—8 eggs in late May, and buries them in the sand. She produces a second clutch in July or August. The young lizards hatch out after a month and measure about 2.5 cm.

RAT SNAKE
Elaphe schrenckii

The rat snake *Elaphe schrenckii* is found in northern China, Korea, eastern Manchuria and in southern regions of the Soviet Far East. It occurs in several subspecies, and attains a length of 2 m. The rings on its body are yellow or white. The young have only crosswise brown spots. This snake lives in forests densely overgrown with bushes, although it prefers to keep to the forest edges or to clearings. It often establishes itself in gardens near houses. It is an excellent climber searching for birds' nests even in the tallest trees. It preys both on eggs and nestlings, and catches small rodents up to the size of a rat. Unlike other snakes, it never flees when it encounters man, but slowly continues on its way. The female lays 13—30 eggs at some time between mid-July and mid-August. The young usually hatch out in September.

WATER SNAKE
Natrix tigrina

The water snake *Natrix tigrina* is distributed in coastal areas near Khabarovsk, in Korea, north-eastern and eastern China, Taiwan, Hainan and Japan. It can attain a length in excess of 130 cm, but the average length is 110 cm. It lives close to water or in damp situations among dense vegetation, and it also occurs in thin woodland. In the water, it preys mainly on fishes, and it catches frogs on the banks. The female lays 18—22 eggs during the second half of July or August. The young usually hatch out at the end of August and

Long-tailed lizard
Takydromus amurensis

measure 15—17 cm long. When it is threatened, this snake attempts to deter its attacker by raising the front part of its body and spreading its neck like a cobra. It has two large teeth in the back of its upper jaw which it uses to kill its prey. These teeth are particularly effective in piercing frogs which inflate themselves when the snake grabs them. When a man gets bitten incidentally, the deep wound absorbs the snake's saliva and a secretion from the venom gland situated beneath the lips. This can result in serious blood poisoning, the symptoms being similar to those in persons bitten by a viper. However, this snake usually inflicts a wound only with its short front teeth. The species does not rank among poisonous snakes because it has no venom fangs.

Rat snake Elaphe schrenckii

Water snake Natrix tigrina

Siberian Salamander

SIBERIAN SALAMANDER
Hynobius keyserlingii

The Siberian Salamander is widespread from the Urals to Kamchatka, Mongolia, North Korea, Sakhalin, the Kuril Islands and Hokkaido. It reaches a length of 13 cm. This salamander lives predominantly in the taiga, usually in damp localities in dense vegetation where it preys on small worms, molluscs and insects. From April to June it lives in water, where the female lays her eggs on aquatic plants. The eggs are produced in a gelatinous sac about 15 cm long and 2 cm wide, each sac containing 50—60 eggs. The newly hatched larvae are 10 mm long and they remain in the water until they are 3—4 cm long, and metamorphosis is complete. In the Tomsk area and on Sakhalin, the larvae spend the winter in the water, leaving it to become land-dwellers in spring.

In 1963, a specimen of this species was excavated in Siberia together with the frozen body of a mammoth. Experts estimated that the mammoth had been entombed in ice for 5 000 years, and laymen believed that the salamander had been there for the same length of time. Unlike the mammoth, the amphibian was found to be still alive. These salamanders hibernate buried deep in the frozen ground, where they show no signs of life, but they awaken with the coming of spring.

ORIENTAL FIRE-BELLIED TOAD
Bombina orientalis

The Oriental Fire-bellied Toad, which is about 5 cm long, is found in north-eastern China, Korea, and in the Soviet Far East. In spring and summer it can be found in lakes, pools, and even puddles in the forests, but as autumn approaches it crawls onto the land and hides away under heaps of leaves in the daytime. In October it takes refuge in a hole among the rocks or beneath the leaves, and hibernates until April. In May, the frogs return to the water, and the females lay 120—200 eggs in clusters of 15—30. The eggs hatch into tadpoles within a few days. They complete metamorphosis by the end of August, and leave the water as tiny frogs. This species feeds on small insects, worms and molluscs.

Oriental Fire-bellied Toad

INDEX

INDEX OF COMMON NAMES

page numbers in italics refer to illustrations

INDEX OF LATIN NAMES

1
2
3
4
5
6
7
8
9
10
11
12
13
14
15